C6X-BASED DIGITAL SIGNAL PROCESSING

Nasser Kehtarnavaz
Burc Simsek

Department of Electrical Engineering
Texas A&M University

Prentice Hall
Upper Saddle River, NJ 07458

Library of Congress Cataloging-in-Publication Data

Kehtarnavaz, Nasser
    Digital Signal Processing / Nasser Kehtarnavaz, Burc Simsek.
      p.   cm. [on file]
    Includes bibliographical references and index.
    ISBN 0-13-088310-7

*To my family, Latifeh,
Nina, and Melina.
NK*

Acquisitions Editor: *Tom Robbins*
Editorial Assistant: *Jessica Power*
Vice-president of production and manufacturing ESM: *David W. Riccardi*
Vice-president and editorial director of ECS: *Marcia Horton*
Executive Managing Editor: *Vince O'Brien*
Managing Editor: *David A. George*
Editorial/production supervision: *Lakshmi Balasubramanian*
Manufacturing Buyer: *Pat Brown*
Marketing manager: *Danny Hoyt*
Composition: *Prepare, Inc.*
Art Studio: *PreTEX, Inc.*

© 2000 Prentice Hall
Prentice Hall, Inc.
Upper Saddle River, New Jersey 07458

The author and publisher of this book have used their best efforts in preparing this book. These efforts include the development, research, and testing of the theories and programs to determine their effectiveness. The author and publisher make no warranty of any kind, expressed or implied, with regard to these programs or the documentation contained in this book. The author and publisher shall not be liable in any event for incidental or consequential damages in connection with, or arising out of, the furnishing, performance, or use of these programs.

All rights reserved. No part of this book may be reproduced, in any form or by any means, whithout permission in writing from the publisher

Printed in the United States of America

10 9 8 7 6 5 4 3 2 1

ISBN 0-13-088310-7

Prentice-Hall International (UK) Limited, *London*
Prentice-Hall of Australia Pty. Limited, *Sydney*
Prentice-Hall Canada Inc., *Toronto*
Prentice-Hall Hispanoamericana, S.A., *Mexico*
Prentice-Hall of India Private Limited, *New Delhi*
Prentice-Hall of Japan, Inc., *Tokyo*
Pearson Education Asia Pte. Ltd., *Singapore*
Editora Prentice-Hall do Brasil, Ltda., *Rio de Janeiro*

Contents

Preface   vii
Acknowledgments   ix

1   Introduction   1

    1.1   A/D Conversion   3
    1.2   Required Software/Hardware   6
    1.3   Organization of Chapters   6

2   TMS320C6x Architecture   8

    2.1   CPU Operation (Dot Product Example)   11
    2.2   DSP Pipelined CPU   13
    2.3   Veloci T3   16

3   Software Tools   17

    3.1   Evaluation Module (EVM) Board   19
    3.2   Assembly File   20
         3.2.1   Directives   21
    3.3   Memory Management and Linking   22
         3.3.1   Linking   23
    3.4   Compiler Utility   24
    3.5   Code Initialization   26
         3.5.1   Data Alignment   31

Lab 1: Getting Familiar with the Software Tools   32

Lab 2: Code Composer Studio Tutorial   40

    L2.1   Creating Projects   41
    L2.2   Debugging Programs   43
         L2.2.1   Breakpoints   43
         L2.2.2   Benchmarking   43
         L2.2.3   Probe Points and File I/O   44
         L2.2.4   Graphing   45
         L2.2.5   Data Monitoring   45
    L2.3   Real-Time Analysis   47

4 Sampling   50

     4.1 Sampling on the C6X EVM   52

Lab 3: Audio Signal Sampling   54

     L3.1 Initializing EVM and Codec   55
     L3.2 Interrupt Data Processing   60

5 Fixed-Point vs. Floating-Point   63

     5.1 Q-Format Number Representation on Fixed-point Processors   63
     5.2 Finite Word Length Effects on Fixed-point Processors   66
         5.2.1 Input or A/D Quantization   66
         5.2.2 Finite Word Length Error   68
     5.3 Floating-Point Number Representation   69

Lab 4: Q-Format and Overflow   70

     L4.1 Scaling   71

6 Code Optimization   75

     6.1 Word Wide Optimization   76
     6.2 Software Pipelining   78
         6.2.1 Linear Assembly   79

Lab 5: Real-Time Filtering   80

     L5.1 Designing a FIR Lowpass Filter   80
     L5.2 Filter Implementation on the C62   83
         L5.2.1 Hand Written Software-Pipelined Assembly   90
         L5.2.2 Assembler Optimizer Software-Pipelined Assembly   93
     L5.3 Filter Implementation on the C67   94

7 Frame Processing   96

     7.1 Triple Buffering   96
     7.2 Direct Memory Access   97

Lab 6: Fast Fourier Transform   97

     L6.1 DFT Implementation   100
     L6.2 FFT Implementation   102
     L6.3 Real-Time FFT   103

8	**Circular Buffering**	**106**
	Lab 7: Adaptive Filtering	**107**
	L7.1 Design of IIR Filter 108
L7.2 IIR Filter Implementation 109
L7.3 Adaptive FIR Filter 111	
9	**Application Examples**	**116**
	9.1 Wavelet-Based Denoising 116
9.2 Efficient Wavelet Reconstruction 117	

Appendix A: FIR Filter Program Listing for Lab 5	**121**
Appendix B: FFT Program Listing for Lab 6	**131**
Appendix C: Adaptive Filter Program Listing for Lab 7	**140**
Appendix D: Quick Reference Guide	**151**
Bibliography	**162**
Index	**163**

Preface

Digital Signal Processing (DSP) has experienced an enormous growth in the last 20 years. Nowadays DSP processors are used in a wide variety of products from cellular phones to motor drives. These processors are expected to play a major role in the next generation of high-speed communication networks. The TMS320C6x processor family has been introduced by Texas Instruments to meet such high-performance demands.

This book has evolved from teaching a DSP laboratory course at Texas A&M University. The objective of this book is to provide the know-how for the implementation and optimization of computationally intensive signal processing algorithms on TMS320C6x DSP processors. It is intended to be used as a textbook for a real-time DSP laboratory course based on the TMS320C6x processor. Such a course is meant to be a follow-up to a first course in DSP. The material presented in this book is primarily written for those who are already familiar with DSP concepts and are interested in real-time and efficient algorithm implementation on the TMS320C6x.

The thrust of this book is not the theoretical signal processing that is covered in very many textbooks, rather it is written to allow the reader to implement signal processing algorithms and examine their performance on the C6x DSP platform. Note that most of the information in this book appears in the TI reference manuals on TMS320C6x [1–7]. However, this information has been restructured, condensed, and modified to be used for teaching a DSP laboratory course based on the TMS320C6x. It is recommended that these manuals are used in conjunction with this book to make full use of the information presented.

Seven lab exercises are discussed and included on an attached CD to take the reader through the entire process of C6x code writing and optimization. As a result, the book can be used as a self-study guide for implementation of signal processing algorithms on the C6x. The chapters are organized to create a close correlation between the topics and lab exercises if they are used as lecture materials for a lab course. Knowledge of the C programming language is required for understanding and performing the lab exercises.

Every attempt has been made to ensure the correctness of codes. We would appreciate readers bringing to our attention (*kehtar@ee.tamu.edu*) any coding errors that may appear in the printed version.

Nasser Kehtarnavaz

Acknowledgments

Foremost, we would like to thank Mr. Gene Frantz at Texas Instruments who suggested writing this book. This book would not have materialized without his encouragement. We would also like to acknowledge Gene Frantz, Allison Frantz, Richard Scales, Chad Courtney, and Michael Shust at Texas Instruments and Dr. Mansour Keramat at University of Connecticut, who reviewed this book and provided useful comments. Thanks are extended to Dr. Chanan Singh at Texas A&M University and Ms. Maria Ho at Texas Instruments, who provided the necessary administrative support to expedite the writing of this book. Finally, we would like to thank the students who provided feedback on the lab exercises, in particular, Sooncheol Baeg, Javier Davila, Kun Zhao, and Yan Jin.

CHAPTER 1

Introduction

Digital signal processing (DSP) involves the manipulation of digital signals in order to extract useful information from them. A digital signal is formed by sampling and quantizing an analog signal. The digitization process is achieved via an analog-to-digital (A/D) converter. The input/output transfer function of A/D converters is shown in Figure 1–1. Basically, an A/D converter converts analog voltage values into discrete voltage values. Many transducers and sensors generate analog signals. Therefore, some sort of an A/D converter is normally needed at the front-end of a DSP system, as is shown in Figure 1–2.

There are many reasons why one would want to process an analog signal in a digital fashion by converting it into a digital signal. The main reason is that digital processing allows programming flexibility. The same DSP hardware can be used for many different applications by simply changing the code residing in memory. Another reason is that digital circuits provide a more stable and tolerant output as compared with analog circuits, for example, when subjected to temperature changes.

In addition, the advantage of operating in the digital domain could be intrinsic. For example, a linear-phase filter can only be designed by using digital signal processing techniques, and many adaptive systems are achievable only via digital manipulation of signals. In essence, digital representation (0s and 1s) allows voice, audio, and video data

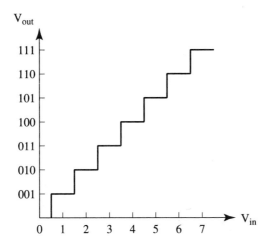

FIGURE 1–1 Three-bit A/D converter transfer function.

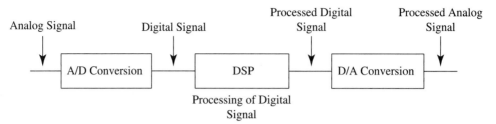

FIGURE 1–2 DSP system main components.

to be treated the same for transmission and storage purposes. As a result, digital processing, and hence digital signal processors (also called DSPs) are expected to play a major role in the next generation of high-speed communication links, including cable (cable modem) and telephone lines (Digital Subscriber Lines, or DSLs).

The processing of a digital signal can be implemented on various platforms, such as a DSP, a customized Very Large Scale Integrated (VLSI) circuit, or a general-purpose microprocessor. Most of the market share of DSPs belong to real-time cost-effective embedded systems, e.g., cellular phones, modems, and disk drives. There are two aspects of real-time processing: (a) input/output sampling rate, and (b) system latencies (delays). Typical sampling rates and latencies for several different applications appear in Table 1–1.

TABLE 1–1 Typical sampling rates and latencies for select applications

Application	I/O Sampling Rate	Latency
Instrumentation	1 Hz	*system dependent
Control	>0.1KHz	*system dependent
Voice	8KHz	<50ms
Audio	44.1KHz	*<50ms
Video	1–14MHz	*<50ms

* Often, a signal may not need to be concerned with latency; for example, TV signal is more dependent on synchronization with audio than the latency. In each of these cases, the latency is dependent on the application.

Some of the differences of DSP implementation as compared with single-function VLSI implementation [e.g., Application Specific Integrated Circuit (ASIC)] are the following:

1. There is a fair amount of application flexibility associated with DSP implementation, since the same DSP hardware can be utilized for different applications. In other words, DSPs are programmable. This is not the case for a hardwired digital circuit.
2. DSP processors are cost-effective due to the fact that they are mass produced. For many applications, a customized VLSI chip normally gets built for a single application and a specific customer.
3. Often very high sampling rates can be achieved by a customized chip, whereas there are sampling rate limitations associated with DSP chips due to peripheral constraints and design of the architecture.

4. In many situations, new features constitute a software upgrade on a DSP instead of requiring new hardware. In addition, bug fixes are normally easier to do for a DSP.

DSPs share some common characteristics that also separate them from general-purpose microprocessors. Some of these characteristics are as follows:

1. With a DSP, it is possible to do several accesses to memory in a single instruction cycle. In other words, DSPs have a relatively high bandwidth between their central processing units (CPUs) and memory.
2. They are optimized to cope with repetition or looping of operations common in signal processing applications.
3. DSPs allow specialized addressing modes, such as indirect and circular addressing. These are efficient addressing mechanisms to implement many signal processing algorithms.
4. They possess appropriate peripherals that allow for efficient input/output (I/O) interfacing.

It should be kept in mind that due to the constant evolving of features being placed on processors, one needs to be cautious of features dividing DSPs and general-purpose microprocessors.

1.1 A/D CONVERSION

Since a DSP needs to interact with the real world made up of analog signals (i.e., input data are normally available in the form of an analog signal from a sensor), some sort of signal conditioning and conversion are done at the front-end of DSP systems. As illustrated in Figure 1–3, signal conditioning encompasses (a) linearization to linearize analog signals as generated by signal sources or sensors, (b) amplification to amplify signals to a level that falls within the analog-to-digital converter range, and (c) bandwidth limitation to limit the highest frequency for antialiasing purposes.

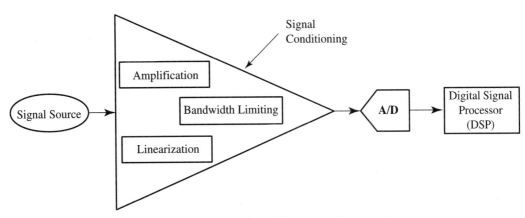

FIGURE 1–3   Signal conditioning and A/D conversion.

4 C6x-Based DSP

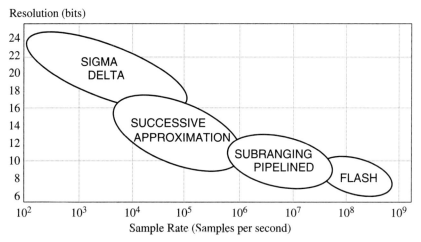

FIGURE 1–4 Resolution vs. sampling rate of ADCs.

There are many types of A/D converters (ADCs) in the market. They include flash, subranging pipelined, successive approximation, and sigma-delta. Figure 1–4 illustrates resolution vs. sampling rate for different types of ADCs.

ADCs can be grouped into two categories: Nyquist rate sampling and oversampling. Here we provide an overview of two popular types: flash (Nyquist sampling) and sigma-delta (oversampling). The interested reader is referred to [9, 10] for more details on ADCs.

Figure 1–5 illustrates a two-bit flash ADC. The input signal voltage V_{in} is compared with a series of reference voltages created by a ladder of laser-trimmed, equal-value resistors. After comparing the reference voltages with the signal voltage via comparators, the difference voltages are decoded into a digital signal. Although the sig-

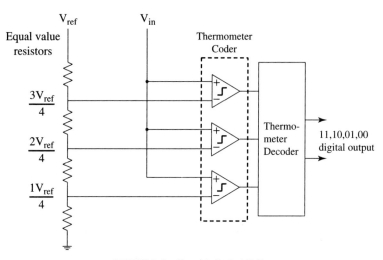

FIGURE 1–5 Two-bit flash ADC.

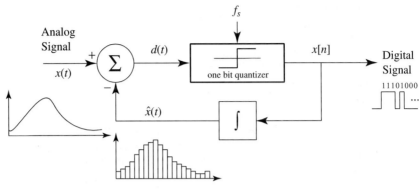

FIGURE 1–6 Delta modulation.

nal conversion is done fast by flash ADCs, its resolution is limited due to the cost involved in getting accurately matched resistor values and having a large number of comparators (2^N, where N is the number of bits). This leads to a large chip area and high power consumption.

An oversampling ADC that has become popular is sigma-delta. There are several advantages associated with this converter. First, for high-resolution conversion they are cost-effective, since they do not require the use of accurate components; that is, they are tolerant to manufacturing process fluctuations. Second, flash and other analog-circuit-based ADCs operate at the Nyquist sampling rate. Sigma-delta operates at sampling frequencies much higher (e.g., 50 times higher) than the Nyquist rate. Consequently, the noise power is considerably reduced, and the design of the front-end antialising filter is made easier.

The sigma-delta converter is a modification of delta modulation used in data compression. To get an idea how such a converter works, consider the block diagram in Figure 1–6 illustrating the delta modulation scheme. Here the difference or error $d(t)$ between the analog signal $x(t)$ and its estimated or integrated version $\hat{x}(t)$ is quantized by a one-bit quantizer. The output signal $x[n]$ consists of zeros and ones corresponding to the one-bit quantized version of the difference signal $d(t)$. The sigma-delta converter, a modification of delta modulation, is shown in Figure 1–7. This modification makes

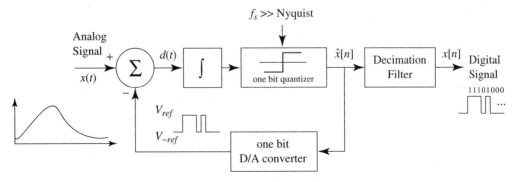

FIGURE 1–7 Sigma-delta conversion.

the conversion process more robust to bit errors that can occur in the converted digital signal. The one-bit oversampled output signal $\hat{x}[n]$ is passed through a decimation filter. This filter averages the bits to produce an N-bit digital signal at a frequency N times lower than the sampling frequency f_s.

1.2 REQUIRED SOFTWARE/HARDWARE

The chapters that follow discuss the implementation and optimization of algorithms to process digital signals on the TMS320C6x DSP. The software tools needed to generate TMS320C6x executable files include the C6x assembler, linker, and compiler. In the absence of an Evaluation Module (EVM) board that allows one to run an executable file on an actual C6x processor, the C6x simulator can be used to verify code functionality by using data already stored in a datafile. However, when using the simulator, an interrupt service routine (ISR) cannot be used to read in signal samples from a signal source. To be able to process digital signals in real-time on an actual C6x processor, an EVM board is needed for code development. The debugger is the software tool required for loading, running, and debugging executable files on an EVM board.

An EVM board can be easily installed in a full-length PCI slot inside a Pentium PC host. Refer to the TI TMS320C6x Evaluation Module Reference Guide [2] for the installation of the board. The PC host ought to have a minimum of 16 MB RAM and be able to accommodate at least 47 MB for the software tools on its hard drive. The interfacing with the board is done through three standard audio jacks appearing at the back of the EVM board. The interfacing equipment may consist of a function generator, oscilloscope, microphone, boombox, and cables with audio jacks.

In addition to the software tools just mentioned, the Code Composer Studio utility can be used to have an easy-to-use graphical interfacing environment on the PC host. For doing the filtering lab assignments, familiarity with MATLAB is helpful, as the filter designs are done by using MATLAB functions; no other filtering package is used.

1.3 ORGANIZATION OF CHAPTERS

We start with an overview of the TMS320C6x architecture in Chapter 2. The focus here is placed on the architectural features one needs to be aware of for implementing algorithms. In Chapter 3, the software tools are presented, and various steps in taking a source file to an executable file are discussed. Lab 1, included in this chapter, provides a hands-on approach for becoming familiar with these tools. This chapter also includes Lab 2 covering the utilization of the Code Composer Studio utility. Chapter 4 presents the process of sampling. Lab 3 in this chapter shows how to sample an input analog signal in real-time on the C6x EVM board. This lab provides the shell program for the lab exercises in the remaining chapters. In Chapter 5, fixed-point and floating-point number representations are presented, and their differences are pointed out. In addition, the finite word length effect is discussed. Lab 4, as part of this chapter, shows how one may cope with the overflow problem. Code efficiency issues appear in Chapter 6 where optimization techniques, as well as linear assembly, are discussed. A finite impulse response (FIR) filtering program is then presented as Lab 5 in this chapter to illustrate different timing cycles for various implementation versions of an algorithm. The idea of frame processing is stated in Chapter 7. As an example of frame processing, Lab 6, included in this chapter, addresses the implementation of the fast Fourier transform (FFT) algorithm and the use of DMA (Direct Memory Access). Lab 7 in Chapter 8 focuses on adaptive filtering and the use of circular buffering. Finally, in Chapter 9, two application examples are included to further illustrate the difference in timing cycles for different implementations of an algorithm. The complete listing of Lab 5, Lab 6, and Lab 7 programs are provided in Appendix A, B, and C, respectively. In addition, a quick reference guide is included in Appendix D at the end of the book.

CHAPTER 2

TMS320C6x Architecture

The choice of a DSP to implement an algorithm in real-time is application dependent. Real-time basically means doing the processing within the allowable time between samples. But beyond real-time, there are many factors that influence this choice. These factors include cost, performance, power consumption, ease of use, time to market, and integration/interfacing capabilities.

The family of TMS320C6x processors, manufactured by Texas Instruments, are built to deliver speed. There are many versions of this processor family having different size, cost, memory, peripherals, and power consumption specifications. At the time of this writing, the fixed-point C6201 version can operate at 200 MHz (5-ns cycle time), delivering a peak performance of 1600 million instructions per second (MIPS), or 400 million multiply and accumulates (MACs) per second. The floating-point C6701 version can operate at 167 MHz (6-ns cycle time), delivering a peak performance of 1000 million floating-point operations per second (MFLOPS). Figure 2–1 illustrates the processing power of C6x by showing a speed benchmarking comparison with some other common DSPs.

Figure 2–2 shows the block diagram of a typical C6x processor. The CPU consists of eight functional units divided into two sides A and B. Each side has a so-called .M unit

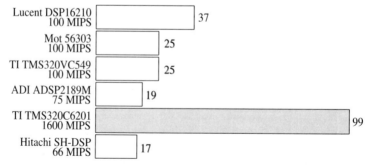

FIGURE 2–1   BDTImark™ Speed Metric benchmark by Berkeley Design Technology, Inc.[1]

---

[1] The BDTImark is a summary measure of DSP speed, distilled from a suite of DSP benchmarks developed and independently verified by Berkeley Design Technology, Inc. A higher BDTImark score indicates a faster processor. For a complete description of the BDTImark and underlying benchmarking methodology, as well as additional BDTImark scores, visit *http://www.bdti.com*. © 1999 Berkeley Design Technology, Inc.

Chapter 2   TMS320C6x Architecture

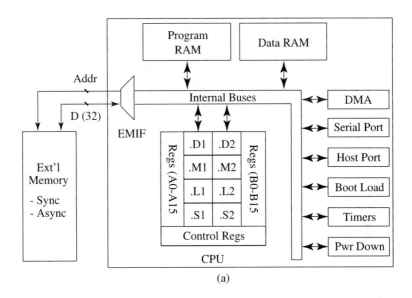

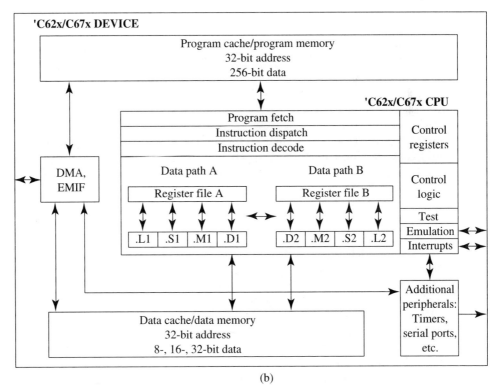

FIGURE 2–2   (a) C6x architecture overview, (b) more detailed block diagram.†

---

†Throughout the book, the figures indicated by † have been reprinted from the TI C6x manuals, courtesy of Texas Instruments.

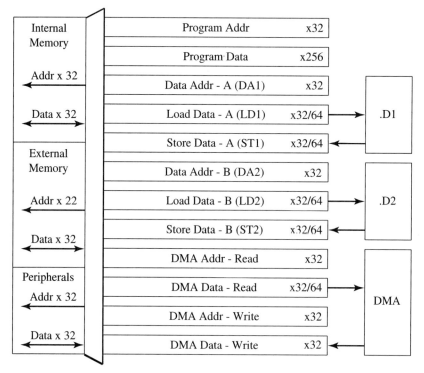

FIGURE 2–3 C6x internal buses.

(used for multiplication operation), a .L unit (used for logical and arithmetic operations), a .S unit (used for branch, bit manipulation, and arithmetic operations), and a .D unit (used for loading/storing and arithmetic operations). Some instructions, such as ADD, can be done by more than one unit. There are sixteen 32-bit registers associated with each side. Interaction with the CPU must be done through these registers. A listing of the C6x instructions as divided by the four functional units appears in Appendix D (Quick Reference Guide). These instructions are fully discussed in the TI TMS320C6x CPU and Instruction Set Reference Guide [1].

As shown in Figure 2–3, the internal buses consist of a 32-bit program address bus, a 256-bit program data bus accommodating eight 32-bit instructions, two 32-bit data address (DA1 and DA2), two 32-bit (64-bit for floating-point version) load data buses (LD1 and LD2), and two 32-bit (64-bit for floating-point version) store data buses (ST1 and ST2). In addition, there are two 32-bit DMA data and two 32-bit DMA address buses. The off-chip, or external, memory is accessed through a 22-bit address and a 32-bit data bus.

The peripherals on the C6201 processor, shown in Figure 2–4, include EMIF (External Memory Interface), DMA, Boot Loader, McBSP (Multi-channel Buffered Serial Port), HPI (Host Port Interface), Timer, and Power Down units. The EMIF provides the necessary timing for accessing external memory. The DMA allows data to move from one place in memory to another place without interfering with the CPU operation. The Boot Loader boots the loading of code from off-chip memory or the HPI to internal memory. The McBSP provides a high-speed multichannel serial communication link.

Chapter 2 TMS320C6x Architecture 11

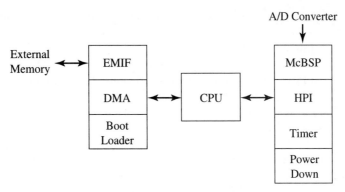

FIGURE 2–4 C6201 peripherals.

The HPI allows a host to access internal memory. The Timer provides two 32-bit counters. The Power Down unit is used to save power for durations when CPU is inactive.

2.1 CPU OPERATION (DOT PRODUCT EXAMPLE)

An effective way to understand the CPU operation is by going through an example. Figure 2–5 shows the code for a 40-point dot product y between two vectors **a** and **x**, $y = \sum_{n=1}^{40} a_n * x_n$. This code appears in the TI Technical Training Notes on TMS320C6x DSP [11].

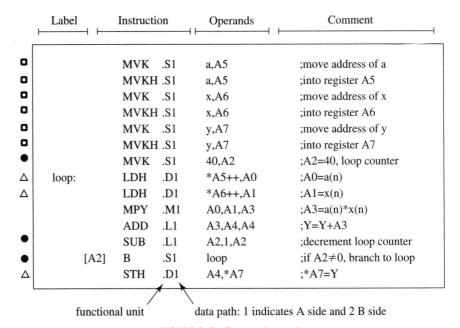

FIGURE 2–5 Dot product code.

The registers assigned to **a**, **x**, loop count, product, y, &a[n] (address of a_n), &x[n] (address of x_n), and &y (address of y) are shown in Figure 2–6. In this example, only the A side functional units and registers are used.

A loop is created by the instructions indicated by ●'s. First, a loop counter is set up by using the move constant instruction MVK. This instruction uses the .S1 unit to place the constant 40 in register A2. The beginning of the loop is indicated by the label *loop* and the end by a subtract instruction SUB to decrement the *loop* counter followed by a branch instruction B to return to *loop*.

The subtraction is done by the .L1 unit and branching by the .S1 unit. The brackets as part of the branch instruction indicate that this is a conditional instruction. All C6x's instructions can be made conditional based on a zero or nonzero value in one of these registers: A1, A2, B0, B1, B2. The syntax [A2] means execute the instruction if A2 ≠ 0, and [!A2] means execute the instruction if A2 = 0. As a result of these instructions, the loop is repeated 40 times.

To start the loop, it is required to set up the A side registers considering that the interaction with the functional units is done through these registers. The instructions labeled by □'s indicate the necessary instructions for doing so. MVK and MVKH are used to load the addresses of a_n, x_n, and y into registers A5, A6, and A7. These instructions must be done in the order indicated to load the lower 16 bits of the full 32-bit address first and then the upper 16 bits. These registers are then used as pointers to load a_n, x_n, into A0, A1 registers and store y from A4 register (instructions labeled by Δ). The C programming language notation * is used to indicate that a register is being used as a pointer. Depending on the datatype, bytes (8-bit) LDB, halfwords (16-bit) LDH, or words (32-bit) LDW loading instructions can be used. Here the data is assumed to be halfwords. The loading/storing is done by the .D1 unit, since .D units are the only units capable of interacting with data memory.

Note that the pointers A5 and A6 need to be post incremented (C notation), so that they point to next values for next iteration of the loop. When registers are used as pointers, there are various ways of performing pointer arithmetic. These include pre-and

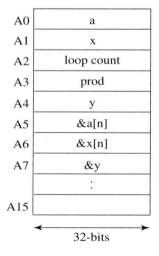

FIGURE 2–6 A-side registers.

Syntax	Description	Pointer Modified
*R	Pointer	No
*+R[disp]	+ Preoffset	No
*−R[disp]	− Preoffset	No
*++R[disp]	Preincrement	Yes
*−−R[disp]	Predecrement	Yes
*R++[disp]	Postincrement	Yes
*R−−[disp]	Postdecrement	Yes

- [disp] specifies # elements - size in W, H, or B
  [disp] = R or 5-bit constant
- (disp) specifies # bytes
  (disp) = 5-bit constant
  +B14/B15(disp) allows 15-bit constant
  Assembler converts to element size [ ]

FIGURE 2–7 Pointer operations.[†]

postincrement/decrement options by some displacement amount where the pointer is modified before or after it is used (e.g., *++A1[disp] and *A1++[disp]). In addition, a preoffset option can be done with no modification of the pointer (e.g., *+A1[disp]). These options are listed in Figure 2–7. At this point, it is worth pointing out that the assembler is not case sensitive; that is, instructions and registers can be written in lowercase or uppercase.

Finally, the instructions MPY and ADD within the loop perform the dot product operation. The instruction MPY is done by the .M1 unit and ADD by the .L1 unit. It should be mentioned that the preceding code, used as is, will not work on the C6x, due to the C6x's unprotected pipeline nature, which is discussed next.

2.2 DSP PIPELINED CPU

In general, it takes several steps to do an instruction, which are basically fetching, decoding, and execution. If these steps are done serially, not all of the resources on the processor, such as multiple buses or functional units, are fully utilized. To increase throughput, DSP CPUs are designed to be pipelined. This means that the foregoing steps are carried out at the same time. A simple example is shown in Figure 2–8 to illustrate the difference in processing time for three instructions executed on a serial/nonpipelined and a pipelined CPU. As can be seen, a pipelined CPU requires fewer clock cycles to do the same instructions.

On the C6x processor, fetching consists of four phases, each requiring a clock cycle. These include generate fetch address denoted by F1, send address to memory F2, wait for data F3, and read opcode from memory F4. Decoding consists of two phases, each requiring a clock cycle. These are dispatching to appropriate functional units denoted by D1 and decoding D2. Due to the delays associated with the instructions multiply

14 C6x-Based DSP

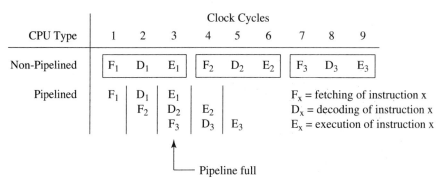

FIGURE 2–8 Pipelined vs. Nonpipelined CPU.†

(MPY-1 delay), load (LDx – 4 delays), and branch (B – 5 delays), the execution step may consist of up to six phases, E1 through E6, accommodating a maximum of 5 delays. In fact, as shown in Figure 2–9, the F step consists of four, the D step of two, and the E step of six substeps or phases.

When the outcome of an instruction is used by the next instruction, an appropriate number of NOPs (no operation or delay) must be added after multiply (one NOP), load (four NOPs, or NOP 4), and branch (five NOPs, or NOP 5) instructions in order to allow the pipeline to operate properly. Therefore, for the previous example to run on the C6x, NOPs, as shown in Figure 2–10, should be added after the instructions MPY, LDH, and B.

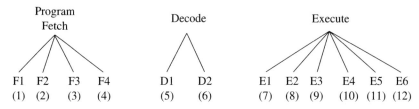

FIGURE 2–9 Stages of the pipeline.

```
           MVK    .S1    40,A2
    loop:  LDH    .D1    *A5++,A0
           LDH    .D1    *A6++,A1
           NOP            4
           MPY    .M1    A0,A1,A3
           NOP
           ADD    .L1    A3,A4,A4
           SUB    .L1    A2,1,A2
    [A2]   B      .S1    loop
           NOP            5
           STH    .D1    A4,*A7
```

FIGURE 2–10 Pipelined code with NOPs inserted.

Figure 2–11 illustrates an example of a pipeline situation that requires adding a NOP. The + signs indicate the number of substeps or latencies required for the instruction to be completed. In this example, it is assumed that the addition operation is done before one of its operands is made available from the previous multiply operation. Hence, the need for adding a NOP after the MPY. Later on, it will be seen that as part of

Prog Fetch	Decode	Execute						Done
F1–F4	D1–D2	E1	E2	E3	E4	E5	E6	
MPY								

Multiply is fetched.

	MPY							
ADD								

Multiply is decoded and ADD is fetched.

		MPY	+					
	ADD							

Multiply is executed and ADD is decoded.

			MPY					
		ADD						

Multiply is still being executed while ADD is also executed.

								➤ MPY
								➤ ADD

Both instructions finish at the same time, the result from the MPY is not used in the ADD instruction.

(a)

Prog Fetch	Decode	Execute						Done
F1–F4	D1–D2	E1	E2	E3	E4	E5	E6	
MPY								

MPY is fetched.

	MPY							
NOP								

MPY is decoded and NOP is fetched.

		MPY	+					
	NOP							
ADD								

MPY is executed, NOP is decoded, and ADD is fetched.

			MPY					
		NOP						
	ADD							

MPY is still being executed while NOP stalls the pipeline, and ADD is decoded.

								➤ MPY
			ADD					

MPY completes and ADD is executed while using the result from the MPY.

								➤ ADD

Add completes.

(b)

FIGURE 2–11 (a) Multiply then add, (b) need for NOP insertion.

16 C6x-Based DSP

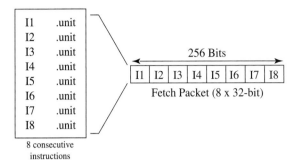

FIGURE 2–12 C6x fetch packet: C6x fetches eight 32-bit instructions every cycle.†

code optimization NOPs can be reduced or removed, leading to an improvement in efficiency.

2.3 VELOCI TI

The C6x architecture is based on the Very Large Instruction Word (VLIW) architecture. In such an architecture, several instructions are captured and processed simultaneously. This is referred to as a Fetch Packet (FP). (See Figure 2–12.)

The C6x uses VLIW allowing eight instructions to be captured simultaneously from on-chip memory onto its 256-bit wide program data bus. The original VLIW architecture has been modified on C6x to allow several Execute Packets (EP) to be included within the same Fetch Packet, as shown in Figure 2–13. An EP constitutes a group of parallel instructions. Parallel instructions are indicated by double pipe symbols (||), and, as the name implies, they are executed together or in parallel. This VLIW modification done by TI is called VelociTI. As compared with VLIW, VelociTI reduces code size and increases performance when instructions reside off-chip.

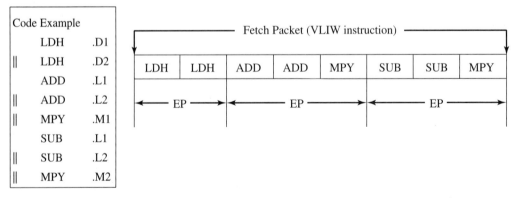

FIGURE 2–13 A fetch packet containing three execute packets.

CHAPTER 3

Software Tools

The programming for most DSPs can be done either in C or assembly. Although writing programs in C would require less effort, the efficiency achieved when doing so is less than that when writing programs in assembly. Efficiency means having as few instructions or as few clock cycles as possible by making maximum use of the resources on the chip.

In practice, one starts with C coding to analyze the behavior and functionality of an algorithm. Then, if the required processing rate is not met by using the C compiler optimizer, the time-consuming portions of the C code are identified and converted into assembly/linear assembly code.

For TMS320 DSPs, the software tools for converting a C or assembly file into a DSP executable file include the compiler, assembler, linker, and debugger/simulator. In addition to C and assembly, the C6x code can be written in so-called linear assembly. Figure 3–1 illustrates the code efficiency vs. coding effort for three types of source files on the C6x: C, linear assembly, and hand-optimized assembly.

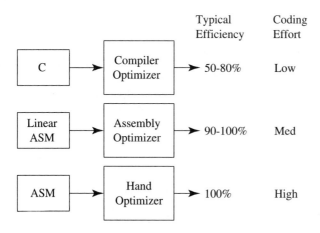

FIGURE 3–1 Code efficiency vs. coding effort.[†]

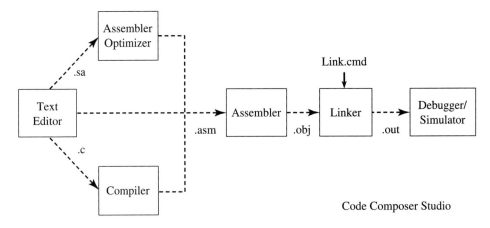

.c = C source file
.sa = linear assembly source file
.asm = assembly source file
.obj = object file
.out = executable file
.cmd = linker command file

FIGURE 3–2 C6x software tools.

As can be seen, linear assembly provides a good compromise between code efficiency and coding effort. Figure 3–2 shows the steps involved for going from a source file (.c extension for C, .asm for assembly, and .sa for linear assembly) to an executable file (.out extension). Figure 3–3 gives the .c and .sa versions of the dot product example to see how they look. The assembler (invoked by the command asm6x) translates an assembly file into an object file (.obj extension). The assembly optimizer and the compiler can be used to translate, respectively, a linear assembly file and a C file into an object file. The linker (invoked by the command lnk6x) combines object files as instructed by the command file (.cmd extension) into an executable file. The simulator (invoked by the command sim62x-sim67x for floating-point version) can then be used to simulate the execution of the executable file on a C62x or C67x processor. A software tool called the Code Composer Studio (CCS)* is also available, providing an easy to use graphical user environment for interfacing and debugging purposes with real-time capabilities.

---

*At the time of writing, TI announced that the simulator and debugger will be made available through the CCS. To provide a more comprehensive approach to the software tools, both versions are covered in the book.

```
main()
{
        y = DotP( (int *) a, (int *) x, 40);
}

int DotP(int *m, int *n, short count)
{
        int sum, i;
        sum = 0;

        for(i=0;i<count;i++)
                sum += m[i] * n[i];

        return(sum);
}
```

(a)

```
                .title "dot product"
                .def dotp
                .sect "code"

dotp:   .proc   A4,B4,A6,B6,A8,B3
                .reg    a, ai, b,bi,r,prod,sum,c,ci,i;
                MV      A4,c
                MV      B4,b
                MV      A6,a
                MV      B6,r
                MV      A8,i

loop:           .trip   40
                LDH     *a++, ai
                LDH     *b++,bi
                MPY     r,bi,prod
                SHR     prod,15,sum
                ADD     ai,sum,ci
                STH     ci, *c++
        [i]     SUB     i,1,i
        [i]     B       loop
                .endproc B3
```

(b)

FIGURE 3-3 (a) .c and (b) .sa version of the dot product example.

3.1 EVALUATION MODULE (EVM) BOARD

Upon the availability of an Evaluation Module (EVM) board, the debugger (invoked by the command **evm6x**) can be activated to run an executable file on an actual C6x processor. The CCS provides a better way to load, run, and debug code on the EVM via

20 C6x-Based DSP

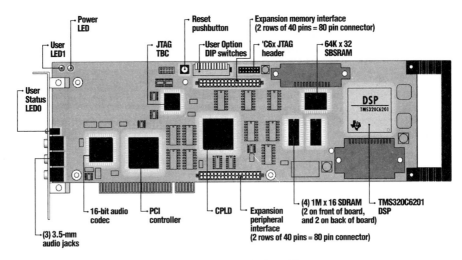

FIGURE 3–4 EVM board.[†]

its graphic user interface (GUI) environment. As shown in Figure 3–4, the C6x EVM board is a complete DSP system that includes a C6x chip, some memory, A/D capabilities, and PC interfacing components. The functional diagram of the EVM board appears in Figure 3–5.

The board has a 16-bit codec that performs the analog-to-digital conversion with sampling frequencies ranging from 5.5 to 48KHz. The memory residing on the EVM board consists of four $1M \times 16$ SDRAM (synchronous dynamic RAM) and one $64K \times 32$ SBSRAM (synchronous burst static RAM), a faster, but more expensive, memory (as compared with SDRAM). A voltage regulator on the board is used to provide 1.8V or 2.5V for the C6x core, 3.3V for its memory and peripherals, and 5V for audio components.

3.2 ASSEMBLY FILE

Similar to other assembly languages, the C6x assembly consists of four fields: label, instruction, operands, and comment. (See Figure 2–5.) The first field is the label field. Labels must start in the first column and must begin with a letter. A label, if present, indicates an assigned name to a specific memory location containing an instruction or data. A menmonic or a directive constitutes the instruction field. It is optional for the instruction field to include the functional unit performing that instruction. However, to make the code more understandable, functional unit assignment is recommended. If a functional unit is specified, the data path must be indexed by 1 for the A side and 2 for the B side. A parallel instruction is indicated by a double pipe symbol (||) and a conditional instruction by a register appearing in brackets in the instruction field. As the name operand implies, the operand field contains arguments of an instruction. Instructions

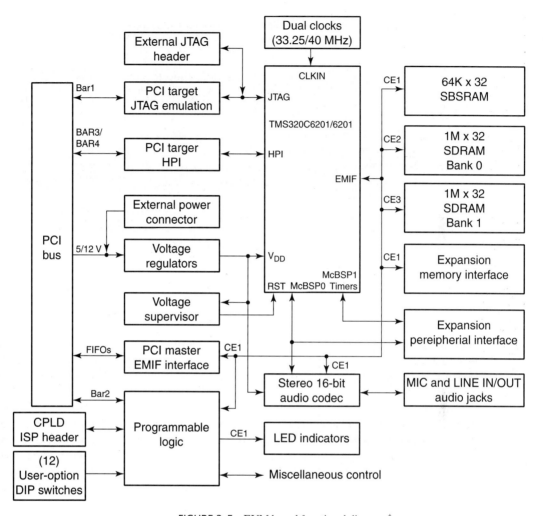

FIGURE 3–5 EVM board functional diagram.[†]

require two or three operands. Except for store instructions, the destination operand must be a register. One of the source operands must be a register, the other a register or a constant. After the operand field, there is an optional comment field that, if stated, should begin with a semicolon (;).

3.2.1 Directives

Directives are used to indicate assembly code sections and to declare data structures. It should be noted that assembly statements appearing as directives do not produce any executable code. They merely control the assemblying process by the assembler. Some of the widely used assembler directives include the following:

22 C6x-Based DSP

.sect "name" directive defining a section of code or data named "name".
.int, .long, or .word directive reserving a 32-bit of memory initialized to a value.
.short or .half directive reserving a 16-bit of memory initialized to a value.
.byte directive reserving an 8-bit of memory initialized to a value.

Note that in the TI Common Object File Format (COFF), the directives .text, .data, and .bss are used to indicate code, initialized constant data, and uninitialized variables, respectively. Other directives often used include .set directive for assigning a value to a symbol, .global or .def directive to declare a symbol or module as global for it to be recognized externally by other modules, and .end directive to signal the termination of assembly code.

At this point, it should be mentioned that the C compiler creates various sections indicated by the .text, .switch, .const, .cinit, .bss, .far, .stack, .sysmem, and .cio directives. For a complete listing of directives, refer to the TI TMS320C6x Assembly Language Tools User's Guide [4].

3.3 MEMORY MANAGEMENT AND LINKING

The external memory used by a DSP can be static or dynamic. Static memory (SRAM) is faster than dynamic memory (DRAM), but it is more expensive, since it takes more space on silicon. DRAMs also require that they are refreshed periodically. A good compromise between cost and performance can be achieved by using SDRAM (Synchronous DRAM). Synchronous memory requires clocking, as compared with asynchronous memory, which does not.

Figure 3–6 illustrates the often-used memory map model-0 for C6201 on the EVM board. As indicated in this figure, there are three 16M external memory ranges numbered

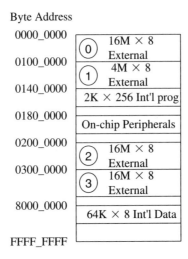

FIGURE 3–6 C6x Memory Map.†

0, 2, and 3, and one 4M external memory range numbered 1. Range 0 starts at 0000_0000, range 1 at 0100_0000, range 2 at 0200_0000, and range 3 at 0300_0000. These external memory ranges support synchronous (SBSRAM and SDRAM) or asynchronous (SRAM, ROM, etc.) memory accessible as bytes (8 bits), halfwords (16 bits), or words (32 bits). There exist 128K bytes of internal RAM, which is equally divided between program and data. The data portion consists of 64K bytes accessible as bytes, halfwords, and words. The program portion consists of 2K fetch packets or 16K 32-bit instructions. The on-chip peripherals and control registers are memory mapped into the memory sections starting at 0140_0000 and 8000_0000, respectively. A listing of the memory mapped registers is provided in Appendix D (Quick Reference Guide). The C6201 on the EVM board can run at a maximum speed of 160 MHz, giving an instruction cycle time of 6 ns.

3.3.1 Linking

Linking places code, constant, and variable sections into appropriate locations in memory as specified in the linker .cmd command file. Also, it combines several object files .obj's into the final executable .out file. A typical command file is shown in Figure 3–7.

```
/* Memory Map 0 */
file1.obj
file2.obj

-o outfile.out

MEMORY
{
    EXT0:      o = 00000000h   l = 01000000h        ; o = origin, l = length
    EXT1:      o = 01000000h   l = 00400000h
    PMEM:      o = 01400000h   l = 00010000h
    EXT2:      o = 02000000h   l = 01000000h
    EXT3:      o = 03000000h   l = 01000000h
    DMEM:      o = 80000000h   l = 00010000h
}

SECTIONS
{
    .text      >   PMEM         ; PMEM = Program memory, DMEM = Data memory
    .stack     >   DMEM         ; EXT = External memory
    .bss       >   DMEM
    .cinit     >   DMEM
    .cio       >   DMEM
    .const     >   DMEM
    .data      >   DMEM
    .switch    >   DMEM
    .sysmem    >   DMEM
    .far       >   EXT2
}
```

FIGURE 3–7 A typical command file.

Section Name	Description
.text	Code
.switch	Tables for switch instructions
.const	Global and static **string literals**
.cinit	Initial values for global/static vars
.bss	Global and static variables
.far	Global and statics declared **far**
.stack	Stack (local variables)
.sysmem	Memory for malloc fcns (heap)
.cio	Buffers for stdio functions

FIGURE 3–8 Common compiler sections.

The first part of the command file indicates the input/output files together with the linking options. The second part (MEMORY) provides a description of the type of physical memory, its origin, and its length. The final part (SECTIONS) specifies the assignment of various code sections to the available physical memory. Figure 3–8 lists some common compiler sections.

3.4 COMPILER UTILITY

It is possible to perform the entire process of compiling, assemblying, and linking in one step by invoking the utility cl6x and stating the right options for it. The following example shows how this utility is used for the source files *file1.c*, *file2.asm*, and *file3.sa*:

cl6x-**gs** *file1.c file2.asm file3.sa* **-z -o** *file.out* **-m** *file.map* **-l** *rts6201.lib*

The option **-g** adds debugger specific information to the object file for debugging purposes. The option **-s** provides an interlisting of C and assembly. For *file1.c*, the C compiler is invoked; for *file2.asm*, the assembler is invoked; and for *file3.sa*, the assembler optimizer (linear assembler) is invoked. The option **–z** invokes the linker placing the executable code in *file.out* if the **-o** option is used; otherwise, the default file *a.out* is created. The option **-m** provides a map file *file.map* that includes a listing of all the addresses of sections, symbols, and labels. The option **-l** specifies the run-time support library *rts6201.lib* for linking files on the C6201 processor. To summarize,

-g generates symbolic debugging directives

-s invokes the interlist utility

-z runs the linker on the specified files

> 1. Compile without optimization.
>   (Get the code functioning!)
>   **cl6x –g –s file.c –z**
>
> 2. Compile with some optimization.
>   (Verify code functionality, again)
>   **cl6x –g –o file.c –z**
>
> 3. Compile with all optimizations.
>   (Generate efficient code)
>   **cl6x –o3 –pm file.c –z**

FIGURE 3–9 Programming approach.

-o specifies the output filename
-m creates a map file
-l specifies a library to be linked

The compiler allows four levels of optimizations invoked by using the **-o0, -o1, -o2**, and **-o3** options. Debugging and full-scale optimization cannot be done together, since they are in conflict; that is, in debugging, information is added to enhance the debugging process, while in optimizing, information is minimized or removed to enhance code efficiency. In essence, the optimizer changes the flow of C code, making the debugging very difficult. For more information on the optimization options, refer to the TI TMS320C6x Optimizing C Compiler User's Guide [5].

As shown in Figure 3–9, a good programming approach would be first to verify that the code is properly functioning by using the compiler with no optimization (**-gs** option). Then, use full optimization to generate an efficient code (**-o3** option). It is recommended to go through an intermediary step where some optimization is done without interfering with source-level debugging (**-go** option). This intermediary step can be done to verify code functionality again before doing full optimization. It should be pointed out that full optimization may change memory locations outside the scope of the C code. Such memory locations must be declared as 'volatile' to prevent compiling errors. Table 3–1 provides some widely used options with the software tools. A useful option listed here is the so-called feedback option (**-mw**). This option can be used to aid the code development process by determining which optimizations might be useful for further improvement in efficiency. Refer to the Optimizing C Compiler manual [5] for more details.

Finally, as a step to further optimize C code, it is recommended that intrinsics be used wherever possible. Intrinsics are those C6x instructions (preceded by an underscore) that are recognized by the C compiler and parsed directly by using C6x instructions. As an example, instead of using the multiply operator * in C, the _mpy() intrinsic can be used to tell the compiler to use the C6x instruction MPY. Figure 3–10 shows the intrinsic version of the dot product C code.

26 C6x-Based DSP

TABLE 3–1 Frequently used options.

Option	Description
-@ *filename*	Appends the contents of a file to the command line.
-c	Suppresses the linker and overrides the –z option, which specifies linking.
-d *name[=def]*	Predefines the constant name for the preprocessor.
-g	Generates symbolic debugging directives that are used by the C source-level debugger and enables assembly source debugging in the assembler.
-I *directory*	Adds *directory* to the list of directories that the compiler searches for #include files.
-k	Retains the assembly language output from the compiler or assembly optimizer.
-me	Produces code in big endian format (little endian is the default).
-ml$n$	Generates large-memory model code on four levels $n = 1, 2, 3, 4$.
-mv6700	Selects the target CPU version as C67x (Default is C62x).
-mw	Enables feedback option.
-n	Compiles or assembly optimizes only.
-q	Suppresses banners and progress information from all the tools.
-qq	Suppresses all output except error messages.
-s	Invokes the interlist utility.
-z	Runs the linker on the specified object files.
-o	Specifies the name of the output file.

```
short DotP(int *m, int *n, short count)
{
        short i, productl, producth, suml = 0, sumh = 0;

        for(i=0;i<count;i++)
        {
                productl = _mpy(m[i],n[i]);          // _mpy intrinsic
                producth = _mpyh(m[i],n[i]);         // _mpyh intrinsic
                suml += productl;
                sumh += producth;
        }
        suml += sumh;
        return(sum);
}
```

FIGURE 3–10 Intrinsic version of the dot product C code.

3.5 CODE INITIALIZATION

All programs start by going through a reset initialization code. Figure 3–11 illustrates both the C and assembly versions of a typical reset initialization code. This initialization is for the purpose of starting at a previously defined initial location. Upon power up, the system always goes to the reset location in memory, which normally includes a branch instruction to the beginning of the code to be executed. The reset code shown in Fig-

```
        "ASM"                                    "C"

vectors.asm                              cvectors.asm

        .ref    init                             .global  _c_int00
        .sect   "vectors"                        .sect    "vectors"
  rst   MVK   .s2    init,B0            rst      B        _c_int00
        MVKH  .s2    init,B0                     NOP              ;additional NOP's
        B     .s2    B0                          NOP              ;to create a
        NOP                                      NOP              ;fetch packet
        NOP                                      NOP
        NOP                                      NOP
        NOP                                      NOP
        NOP                                      NOP
```

FIGURE 3–11 Reset code.

ure 3–11 takes the program counter to a globally defined location in memory named *init* or *_c_int00*.

As shown in Figure 3–12, when writing in assembly, an initialization code is needed to create initialized data and variables and to copy initialized data into corresponding variables. Initialized values are specified by using .byte, .short, or .int directives. Uninitialized variables are specified by using .usect directive. Before calling the main function or subroutine, an initialization code is usually needed to set up registers, pointers, and to move data to appropriate places in memory.

Figure 3–13 provides the initialization code for the dot product example where initialized data values appear for three initialized data arrays labeled *table_a*, *table_x*, and *table_y*. In addition, three variable sections called *a*, *x*, and *y* are declared. The second part of the initialization code copies the initialized data into the corresponding variables. The setup code for calling the dot product routine is also shown in this figure.

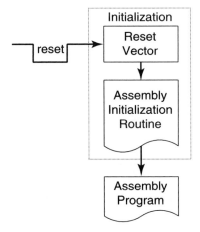

FIGURE 3–12 Assembly initialization.†

```
            .def      init
            .ref      dotp

;Data initialization
;Initialize tables

            .sect     "init_tables"

table_a     .short    40,39,38,37,36,35,34,33,32,31,30,29,28,27    ;Initialize table_a array with values
            .short    26,25,24,23,22,21,20,19,18,17
            .short    16,15,14,13,12,14,10,9,8,7,6,5,4,3,2,1
table_x     .short    1,2,3,4,5,6,7,8,9,10,11,12,13,14,15          ;Initialize table_x array with values
            .short    16,17,18,19,20,21,22,23,24,25,26,27,28,29
            .short    30,31,32,33,34,35,36,37,38,39,40
table_y     .short    0                                            ;table_y = 0

;Variable declaration
a           .usect    "var", 80, 2                                 ;define variables
x           .usect    "var", 80, 2
y           .usect    "var", 2, 2

;Initialization to copy data into variables
            .sect     "init_code"
init        mvk       .s1       table_a, A0                ;move address of table_a to register A0
            mvkh      .s1       table_a, A0
            mvk       .s2       a,B0                       ;move address of a to register B0
            mvkh      .s2       a,B0
            mvk       .s2       40,B1                      ;create a counter in register B1, B1=40
loop_a      ldh       .d1       *A0++,A1                   ;load an element from the address pointed by A0 into A1
            sub       .l2       B1,1,B1                    ;decrement counter
            nop                 3
            sth       .d2       A1,*B0++                   ;store the element to address pointed by B0
   [B1]     b         .s2       loop_a                     ;branch back to loop_a
            nop                 5                          ;required latency
init_x      mvk       .s1       table_x, A0                ;move address of table_x into register A0
            mvkh      .s1       table_x, A0
            mvk       .s2       x, B0                      ;move address of x into register A0
            mvkh      .s2       x, B0
            mvk       .s2       40, B1                     ;create a counter
loop_x      ldh       .d1       *A0++,A1                   ;load the element from the address pointed by A0 into A1
            sub       .l2       B1,1,B1                    ;decrement counter
            nop                 3
            sth       .d2       A1,*B0++                   ;store element to address pointed by B0
   [B1]     b         .s2       loop_x                     ;branch back to loop_x
            nop                 5
init_y      mvk       .s1       table_y, A0                ;repeat above procedure for table_y
            mvkh      .s1       table_y, A0
            mvk       .s2       y, B0
            mvkh      .s2       y, B0
            ldh       .d1       *A0, A1
            nop                 4
            sth       .d2       A1, *B0
```

(a)

FIGURE 3–13 (a) Initialization code for dot product example.

```
;Setup for calling dotp

start   mvk     .s1     a,A4            ;move a into register A4
        mvkh    .s1     a,A4
        mvk     .s2     x,B4            ;move x into register B4
        mvkh    .s2     x,B4
        mvk     .s1     40,A6           ;create a counter in A6, A6=40
        b       .s1     dotp            ;branch to routine dotp
        mvk     .s2     return, B3      ;store return address in B3
        mvkh    .s2     return, B3
        nop             3

;return from dotp here
return  mvk     .s1     y, A0           ;move y into register A0
        mvkh    .s1     y, A0
        sth     .d1     A4, *A0         ;store the result of dotp (returned in A4) to y

;infinite loop
end     b       .s1     end             ;infinite loop
        nop             5
```

(b)

```
;dotp

        .def    dotp

;A4 = &a, B4 = &x, A6 = 40 (iteration count) , B3 = return address

dotp    mv              A6,B0           ;move A6 to B0 (third argument passed from calling function)
        zero            A2              ;zero the sum register A2

loop    ldh     .d1     *A4++,A5        ;load an element from the location pointed by A4 into A5
        ldh     .d2     *B4++,B5        ;load an element from the location pointed by B4 into B5
        nop             4
        mpy     .m1x    A5,B5,A5        ;A5=B5*A5
        nop
        add     .l1     A2,A5,A2        ;A2 += A5
 [B0]   sub     .l2     B0,1,B0         ;decrement counter B0
 [B0]   b       .s1     loop            ;branch back to loop
        nop             5

        mv              A2,A4           ;move result in A2 to return register A4
        b       .s2     B3              ;branch back to calling address stored in B3
        nop             5
```

(c)

FIGURE 3–13   (b) Setup code for calling dot product routine; (c) dot product routine.

30 C6x-Based DSP

```
filename.obj
vectors.obj
-o filename.out
MEMORY
{
        .....
}
SECTIONS
{
        .....
}
```

FIGURE 3–14 Linker command file including vectors.obj.

Note the need for the inclusion of *vectors.obj* in the command file as shown in Figure 3–14. (Command files will be further discussed in Lab 1.)

As far as C coding is concerned, the C compiler uses *boot.c* in the run-time support library to do this initialization before calling main(). The **-c** option activates *boot.c* to autoinitialize variables. This is illustrated in Figure 3–15 and the corresponding command file is shown in Figure 3–16.

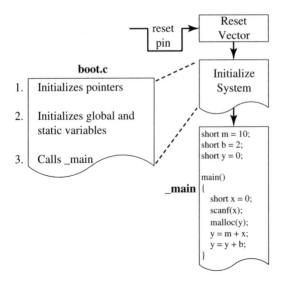

FIGURE 3–15 C initialization.[†]

```
filename.obj
cvectors.obj
-o filename.out
-c                      /*autoinit variables*/
-stack 200h             /*set stack size and heap size*/
-heap 200h              /*default is 400h*/
-l rts6201.lib          /*contains boot.c*/
MEMORY
{
       ....
}
SECTIONS
{
       .....
}
```

FIGURE 3–16 Linker command file including cvectors.obj.

3.5.1 Data Alignment

The C6x allows byte, half-word, and word addressing. Consider the word format representation of memory shown in Figure 3–17. There are four byte boundaries, two half-word or short boundaries, and one word boundary per word. The C6x always accesses data on these boundaries depending on the data type specified; that is, it always accesses aligned data. When specifying an uninitialized variable section .usect, it is required to specify the alignment, as well as the total number of bytes. The examples shown in Figure 3–18 illustrate data alignment for both constants and variables.

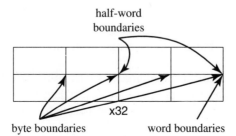

FIGURE 3–17 Data boundaries.

```
                                                                        x32 (le)
┌─────────────────────────────────────────────────────┐   ┌──────────────────────┐
│  Constants are automatically aligned!               │   │  22  22  --  11      │
│         .sect "my_const"                            │   ├──────────────────────┤
│                                                     │   │  33  33  33  33      │
│    A      .byte    11h                              │   ├──────────────────────┤
│    B      .short   2222h                            │   │  --  e   e   d       │
│    C      .int     33333333h                        │   ├──────────────────────┤
│                                                     │   │  f   f   f   f       │
├─────────────────────────────────────────────────────┤   ├──────────────────────┤
│  Variables need an alignment field                  │   │  g1      g0          │
│  ; label    .usect   "section", # bytes, alignment  │   ├──────────────────────┤
│                                                     │   │  g3      g2          │
│    d        .usect   "vars", 1, 1                   │   └──────────────────────┘
│    ee       .usect   "vars", 2    ;byte alignment by default   Note1: Assume that vars and my_const
│    ffff     .usect   "vars", 4, 4                                      sections are contiguous.
│    g_array  .usect   "vars", 8, 2                              Note2: First declare words. Then declare
│                                                                        shorts and bytes to save
└─────────────────────────────────────────────────────┘                  memory space.
```

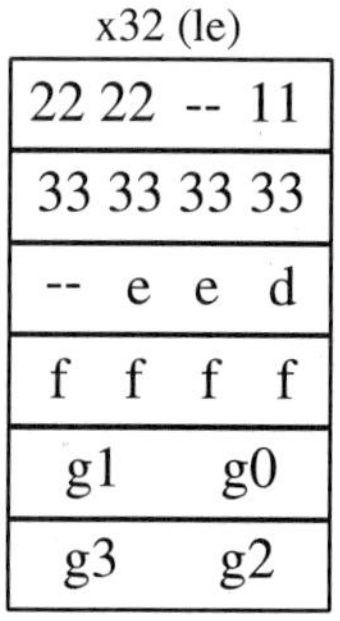

FIGURE 3–18   Constant and variable alignment examples.[†]

Data in memory can be arranged either in little- or big-endian format. Little-endian (le) means that the least significant byte is stored first. Figure 3–19(a) shows storing .int 40302010h in little-endian format for byte, half-word, and word access addressing. In big-endian (be) format, shown in Figure 3–19(b), the most significant byte is stored first. Little-endian is the format normally used in most applications. Additional data alignment examples are shown in Figure 3–19(c) based on the little endian arrangement appearing in Figure 3–19(a).

LAB 1   GETTING FAMILIAR WITH THE SOFTWARE TOOLS

This lab demonstrates how a simple multifile algorithm can be compiled, assembled, and linked by using the software development tools. First, several data values are consecutively written to memory. A pointer is assigned to the beginning of the data so that they can be used in an array. Then, simple functions are added in both C and assembly to illustrate how function calling works. This method of placing data in memory is simple to use and can be used in applications where constants need to be in memory, such as filter coefficients and FFT twiddle factors. Issues related to debugging and benchmarking are also covered in this lab.

We start by creating a command file to use with the programs that follow. This is done by defining memory sections as part of a command file shown in Figure L1–1. This command file, named *link.cmd*, is configured based on the memory map model-0 depicted in Figure 3–6.

Now let us create an assembly file *initmem.asm* by using `.short` directives to declare 10 values from 1 through 10. Note that any of the data-to-memory allocation directives can be used to do this. These values are considered to make up a section named `.sect ".mydata"`; as the following shows:

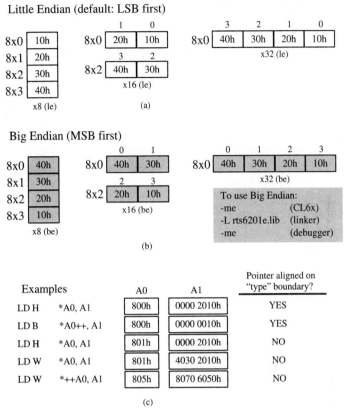

FIGURE 3–19 (a) Little endian, (b) big endian, (c) more data alignment examples.[†]

```
.sect ".mydata"
.short      1
.short      2
.short      3
.short      4
.short      5
.short      6
.short      7
.short      8
.short      9
.short      10
```

To assemble this array of values into the memory, a space that will not be overwritten by the compiler should be selected. The external memory space ranging from EXT1 to EXT3 can be used for this purpose. For now, let us start to assemble the data at the memory address 0x03000000h, which is the beginning of EXT3. In the command file, the section named .mydata is assigned to EXT3 by adding .mydata > EXT3 to the SECTIONS part of the command file as shown in bold in Figure L1–1.

```
/* Memory Map 0 – the default */
MEMORY
{
    EXT0:       o = 00000000h   l = 01000000h
    EXT1:       o = 01000000h   l = 00400000h
    PMEM:       o = 01400000h   l = 00010000h
    EXT2:       o = 02000000h   l = 01000000h
    EXT3:       o = 03000000h   l = 01000000h
    DMEM:       o = 80000000h   l = 00010000h
}


SECTIONS
{
  .text      >   PMEM
  .stack     >   DMEM
  .bss       >   DMEM
  .cinit     >   DMEM
  .cio       >   DMEM
  .const     >   DMEM
  .data      >   DMEM
  .switch    >   DMEM
  .sysmem    >   DMEM
  .far       >   EXT2
  .mydata    >   EXT3

}
```

FIGURE L1–1 Command file for Lab 1.

The file is then assembled by typing the command `asm6x initmem.asm`. To link the file, the following lines are added to the beginning of the command file.

```
initmem.obj
main.obj
-c
-l        rts6201.lib
-o        test.out
```

This command file allows the output *initmem.obj* from the assembler and the output *main.obj* from the compiler to be linked together to create an executable file named *test.out*. The **-c** option allows the linker to use the conventions as required by the C compiler. The **-l** option specifies the use of the run-time support library *rts6201.lib*. The **-o** option makes *test.out* the executable file instead of the default file *a.out*. For debugging purposes, the following empty shell C program is used.

```
#include <stdio.h>
#include <time.h>

void main()
{
        printf("BEGIN\n");
        printf("END\n");
}
```

Chapter 3 Software Tools 35

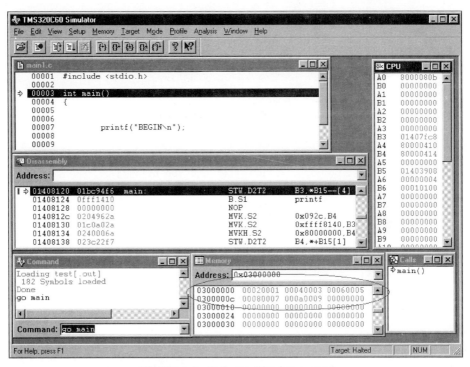

FIGURE L1–2 Debugger/Simulator screen.

Assemble and link this shell program by invoking the assembler and linker as follows:

```
>cl6x -c -g main.c
>lnk6x link.cmd
```

The files should be built without any error. You should be able to see the object files *.obj's* and the *test.out* file. If using earlier versions of the software tools, start the C62x simulator with the commond sim62x. (See Figure L1–2.) In the latest version of the software tools, this simulator/debugger capability can only be activated from the Code Composer Studio, discussed in Lab 2.

To load the file *test.out,* enter the command load test in the command window of the simulator. (Make sure you are in the right directory.) Next, enter go main in the command window. This will take the program to the beginning of the function main(). From the View menu, select mixed option to view all windows. From the Memory window, type 0x03000000 and press enter to view the successful appearance of the data, as circled in Figure L1–2, at the beginning of the EXT3 memory space. Finally, run the program by typing run to see BEGIN and END appearing in the command window.

To assign a pointer to the beginning of the assembled memory space, the memory address can be directly typed in to a pointer. It is necessary to typecast the pointer to a short, since the values are of that type. The following code can be used to assign a pointer to the beginning of the values and loop through them to print each one to the command window:

```
#include <stdio.h>
#include <time.h>

void main()
{
        int i;
        short *point;
        point = (short *) 0x03000000;
        printf("BEGIN\n");
        for(i=0;i<10;i++)
        {
                printf("%d\n",point[i]);
        }
        printf("END\n");
}
```

Build the program and launch the debugger with the command evm6x (for earlier versions of the software tools). Load the program *test.out* and type go main. Before running the program, type wa point and wa i. This should bring up a Watch window where you will see the uninitialized variables. Step through the C program by pressing F8 or by using the Target menu to verify that the program is working.

To add a simple C function that sums the values, we can simply pass the pointer to the array and have a return type of integer. Since we are not using assembly at this point, we are not concerned about how the CPU passes the variables, but rather how much time it takes to perform this operation.

The following simple function can be used to sum the values and return the result:

```
#include <stdio.h>
#include <time.h>

int ret_sum(const short* array, int N);
void main()
{
        int i,ret;
        short *point;
        point = (short *) 0x03000000;
        printf("BEGIN[backslash]n");
        for(i=0;i<10;i++)
        {
                printf("%d\n",point[i]);
        }
        ret = ret_sum(point,10);      <-Breakpoint
        printf("Sum = %d[backslash]n",ret); <- Breakpoint
        printf("END\n");
}
```

```
int ret_sum(const short* array,int N)
{
        int count,sum;
        sum=0;
        for(count=0 ; count < N ; count++)
                sum += array[count];
        return(sum);
}
```

By using this program and running the debugger, we can obtain how much time it takes for the function ret_sum() to run. By adding breakpoints to the line, where this function is called and the next line we can benchmark it. To do such a benchmarking, a breakpoint is set at the calling line, by right clicking on the line and selecting ADD BREAKPOINT from the pop-up menu. Another breakpoint is set at the next line. By typing run, the program runs until the first breakpoint is reached. Now type runb, which enables the variable clk to count the number of cycles it takes until the next breakpoint. By typing wa clk, the number of cycles can be displayed in a Watch window.

Instead of using the debugger, the simulator loader may be used. This tool only simulates the program without running the debugger. By entering load62x test.out at the command line, the program is simulated and the results can be directly viewed at the DOS prompt.

The preceding code is modified so that it includes timing statements, as shown in the next code block. These time functions, declared in the *time.h* header file, can be used to benchmark programs without using the debugger, considering that the debugger does not support them. For additional debugger/simulator commands, refer to the TI TMS320C6x C Source Debugger User's Guide [6]. A summary of the debugger/simulator commands is included in Appendix D (Quick Reference Guide).

```
#include <stdio.h>
#include <time.h>

int ret_sum(const short* array, int N);
void main()
{
        clock_t time_overhead, time_start, time_stop;
        int i,ret;
        short *point;
        point = (short *) 0x03000000;
        for(i=0;i<10;i++)
        {
                printf("%d\n",point[i]);
        }
```

```c
/***************************************************************/
/* COMPUTE THE OVERHEAD OF CALLING CLOCK TWICE TO GET TIMING INFO.    */
/***************************************************************/
time_start    = clock();
    time_stop     = clock();
    time_overhead = time_stop - time_start;
printf("Overhead = %d\n",time_overhead);
/***************************************************************/
/* TIME RET_SUM                                                 */
/***************************************************************/
time_start = clock();
ret = ret_sum(point,10);
time_stop = clock();
printf("RET_SUM: %d cycles, result: %d\n",
        time_stop - time_start - time_overhead,
        ret);
printf("END\n");
}
int ret_sum(const short* array,int N)
{
    int count,sum;
    sum=0;
    for(count=0 ; count < N ; count++)
        sum += array[count];
    return(sum);
}
```

To use an assembly program to calculate the sum of the values, we have to write an assembly file, assemble it separately, and then link the object files to create the executable .out file. Such an assembly file is shown next. Here, the two arguments of the sum function are passed in registers A4 and B4. The return value gets stored in A4 and the return address in B3. Also, the name of the function is preceded by an underscore as .global _sum.

sum.asm

```
        .global   _sum

_sum:
        ZERO      .L1    A9           ;Sum register
        MV        .L1    B4,A2        ;initialize counter with passed argument

loop:   LDH       .D1    *A4++, A7    ;load value pointed by A4 into register A7
        NOP       4
        ADD       .L1    A7,A9,A9     ;A9 += A7
 [A2]   SUB       .L1    A2,1,A2      ;decrement counter
 [A2]   B         .S1    loop         ;branch back to loop
        NOP       5

        MV        .L1    A9,A4        ;move result into return register A4
        B         .S2    B3           ;branch back to address stored in B3
        NOP       5
```

The function `main()` must also be modified by adding a function declaration and a function call. The declaration is done by using the modifier `extern` while not declaring the arguments to the function. This program is shown in Figure L1–3. To build the program, the name of the object file *sum.obj* is added to the command file. The following commands are then entered:

main.c

```
#include <stdio.h>
#include <time.h>

int ret_sum(const short* array, int N);
extern sum();

void main()
{
        clock_t time_overhead, time_start, time_stop;
        int i,ret,asmret;
        short *point;
        point = (short *) 0x03000000;
        printf("BEGIN\n");

        for(i=0;i<10;i++)
        {
                printf("%d\n",point[i]);
        }

        /********************************************************************/
        /* COMPUTE THE OVERHEAD OF CALLING CLOCK TWICE TO GET TIMING INFO*/
        /********************************************************************/

        time_start    = clock();
        time_stop     = clock();
        time_overhead = time_stop - time_start;

        printf("Overhead = %d\n",time_overhead);

        /********************************************************************/
        /* TIME RET_SUM                                                    */
        /********************************************************************/
        time_start = clock();
        ret = ret_sum(point,10);
        time_stop = clock();

        printf("RET_SUM: %d cycles, result: %d \n",
                    time_stop - time_start - time_overhead,
                    ret);

        /********************************************************************/
        /* TIME ASM SUM                                                    */
        /********************************************************************/
        time_start = clock();
        asmret = sum(point,10);
        time_stop = clock();

        printf("ASM_SUM: %d cycles, result: %d \n",
                    time_stop - time_start - time_overhead,
                    asmret);

        printf("END\n");
}

int ret_sum(const short* array,int N)
{
        int count,sum;
        sum=0;
        for(count=0 ; count < N ; count++)
                sum += array[count];

        return(sum);
}
```

FIGURE L1–3 Complete program for Lab 1.

40 C6x-Based DSP

```
>ams6x sum.asm
>cl6x -c -g main.c
>lnk6x link.cmd
>load62x test.out
```

Table L1–1 shows the number of cycles it takes for the program to run using several different build options. As can be seen, the last option corresponding to full optimization is a considerable improvement over the initial build of the program.

TABLE L1–1 Number of cycles for different build options.

Type of Build	Number of Cycles
Standard Build	308
No –g option	160
-o1	99
-o2/3	43

LAB 2 CODE COMPOSER STUDIO TUTORIAL

The Code Composer Studio is a useful utility that allows one to build and debug programs from within a user-friendly GUI. It extends the capabilities of the code development tools to include real-time analysis. Figure L2–1 shows the phases associated with the CCS code development cycle.

The CCS is configurable to operate with the C6201 and C6701 evaluation module boards. It may also be configured to simulate the C6201, C6202, C6203, C6701, and C6711 architectures in big- or little-endian. It includes an integrated code editor that allows the editing of C and assembly source codes. For building applications, the CCS provides an easy-to-use Project Manager window. For debugging purposes, it provides breakpoints, watch variables, view memory/registers/stack, and probe points to stream data to and from the target, graph signals/data, profile execution, and view disassembled and C instructions as they execute on the target. Furthermore, it includes the General Extension Language (GEL) to create custom functions for commonly performed tasks.

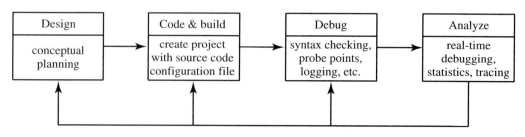

FIGURE L2–1 CCS code development cycle.[†]

The real-time capabilities of the CCS is included in the DSP/BIOS features: Plug-ins and API. The DSP/BIOS Plug-ins allow the user to probe, trace, and monitor a program without disrupting its real-time behavior. The DSP/BIOS API incorporates modules that allow the user to configure global run-time parameters and properties for objects such as software interrupts, I/O pipes, and event logs. The BSP/BIOS API configuration files (.cdb extension) simplify the tasks of memory mapping and ISR vector mapping, and they minimize the target memory footprint by configuring objects statically. The CCS also provides real-time data exchange (RTDX) between the DSP and the host. This allows the user to pass data to and from the target without interrupting the real-time operation of the program.

This short tutorial lab introduces the basic features one needs to know in order to build and debug projects with the Code Composer Studio. For more details, refer to [8].

L2.1 Creating Projects

One important benefit of using the CCS is the ability to create and manage large projects from a GUI environment. The project toolbar, shown in Figure L2–2, which is activated from VIEW/PROJECT TOOLBAR, displays the files included in a project.

Consider all the items required to create an executable file (i.e., .c, .asm, .sa source files, a command file, include and library files, and possibly a .cdb DSP/BIOS configuration file). From the command line, each of these items have to be compiled, assembled, linked, and included separately. From the CCS, the task of including a file to a project is quite easy. Once a new project file (.mak extension) has been created from the PROJECT / NEW menu, several new options appear at the PROJECT menu command. Amongst these is ADD FILES TO PROJECT. Selecting this option allows the user to include source files (.c, .asm, .sa), object files (.obj), library files (.lib), command files (.cmd), and

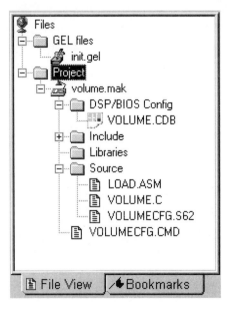

FIGURE L2–2   Project toolbar.

DSP/BIOS configuration files (.cdb) into the project. All the required elements of a project are included in the project toolbar providing a hierarchical view of the project. To peruse through the files within the project, double click the mouse button while the cursor is over the file of interest.

The project settings, which define how the tools behave, are adjusted from the PROJECT/OPTIONS menu. From there, the compiler, assembler, and linker options can be entered, as depicted in Figure L2–3. Once all the required files are in place, the REBUILD ALL command is used to make the executable .out file.

Any error in the build of the project appears in the build window. Figure L2–4 shows a simple "hello" program that is built with a syntax error. The compiler reports the error in red. An error message is printed prompting the user to link the project.

Note that for include files to appear in the project toolbar, a successful build is initially required. Then, all these files will appear in the INCLUDE section of the project toolbar.

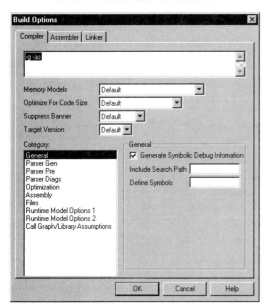

(a)

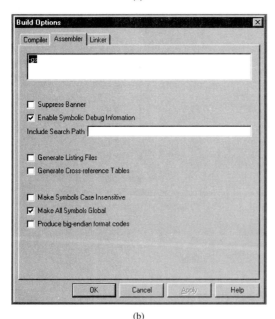

(b)

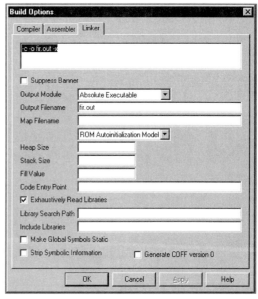

(c)

FIGURE L2–3 (a) Compiler, (b) assembler, and (c) linker project options.

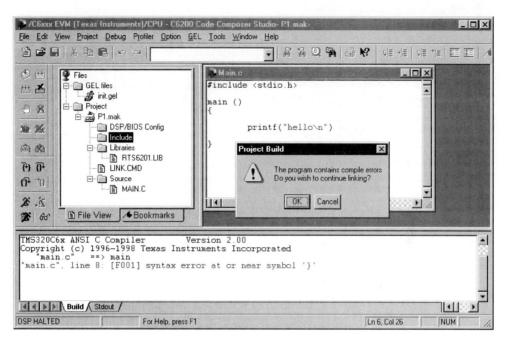

FIGURE L2–4   Build errors.

L2.2 Debugging Programs

The Code Composer Studio allows the user to debug the project by using a variety of standard debugging features, such as breakpoints, runto cursor, and variable watching. Several custom options are available, such as graphing data arrays directly from memory locations, animating processes, and probing files. These are extremely useful in developing C6x code, since most projects start by simulating the response of an algorithm using input and output files.

L2.2.1 Breakpoints   Adding a breakpoint to a specific line is quite simple. The command to enable a breakpoint can be given either from the toggle breakpoint button or by selecting the line with the right mouse button and choosing toggle breakpoint. Once enabled, the line appears as a magenta color to indicate that a breakpoint is set at that line. The program will run up to that line without executing it. Conditional breakpoints can be set using the breakpoint dialog box (i.e., to halt program execution only if the expression is true). From the DEBUG / BREAKPOINTS menu command, individual breakpoints can be enabled, disabled, or deleted.

L2.2.2 Benchmarking   To calculate how many cycles it takes for a particular piece of code to execute, profile points are used in conjunction with breakpoints. Such points are selected similar to breakpoints by using the toggle profile point button. Once this button is selected, a green line appears at the selected location. To do a benchmarking, say, of a function call, two breakpoints and two profile points are required. A breakpoint and a profile point are set on the function call line and again on the line directly after it. Both points would now appear as half green and half magenta, indicating that there are both a breakpoint and a profile point at that line. To enable the profile clock, the menu command PROFILER / ENABLE CLOCK should be selected. The number of cycles can be seen by choosing either VIEW CLOCK or VIEW STATISTICS from the PROFILER menu command. A simpler way is to choose VIEW CLOCK and run the program. The program stops at the first breakpoint at the function call. The line would now have a yellow bar along with a green and magenta color to indicate that the program execution has halted at that line. The clock display window displays how many cycles it takes to get to that line. By double clicking on the clock display, the counter can be reset to zero. By pressing run again, the benchmarking gets started. When the program stops at the next break/profile point, the clock window displays how many cycles it takes to get to that point from the previous point. Notice that this same process can be done without using breakpoints, but instead by using the runto cursor command activated by right clicking on the line. Gathering data on the execution of code is also possible by using the STS module of the DSP / BIOS. (See [8] for details.)

L2.2.3 Probe Points and File I/O Probe points are used to indicate where the program should revert to for action. The file I/O is a useful feature that allows data to be read from a file. This feature probes the file by reading a selected amount of data samples when a probe point is reached, thus simulating a real-time operation, such as an interrupt.

To select a file for input/output, the FILE / FILE I/O menu command should be chosen. This brings up the dialog box shown in Figure L2–5. From there, the input file can be selected and connected to an input address, such as a C symbolic variable (array or variable). The parameter length field defines how many data samples are to be read at every probe point. The wrap-around check box allows the file to be reset when the end is reached.

To select a probe point, the command button should be used. Once this button is selected, a blue line appears indicating the presence of a probe point. One last thing that has to be done is the connection of the probe point to the input/output files. This can be

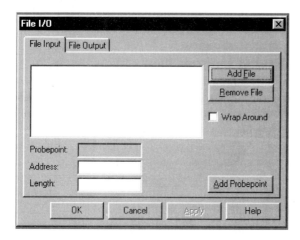

FIGURE L2–5   File I/O dialog box.

FIGURE L2–6 Graph Property Dialog box.

done from the DEBUG / PROBE-POINTS menu where the selected probe point is connected to the input/output files. When the program reaches a probe point, it goes to the connected file and reads the required amount of data to the specified symbolic variable name.

L2.2.4 Graphing The Code Composer Studio allows the user to plot data directly from memory. This can be done by either specifying the memory location and its length or by simply stating the symbolic variable name of an array. The VIEW / GRAPH command allows the user to select the type of plot to be displayed (time/frequency, constellation, eye diagram, image). A simple way to see what is happening in a program is to plot a time graph of data. By selecting VIEW / GRAPH / TIME/FREQUENCY, the Graph Property Dialog box is displayed as shown in Figure L2–6. This Graph Property Dialog box is used to set data parameters such as address, length, type, and Q format.

L2.2.5 Data Monitoring The CCS allows the user to view the contents of memory locations, registers, symbolic variables and the stack. To view the contents of a specific memory location, the VIEW / MEMORY menu command is used. A memory dialog box is used to set the memory location and the viewing format. Figure L2–7 shows the Memory Window dialog box along with the resulting watch window. From the Memory Window, several viewing formats can be selected including hexadecimal, integer, float, exponent, and character.

The contents of the CPU registers, peripheral registers, DMA registers, and serial port registers can also be viewed. Selecting VIEW / CPU REGISTERS will prompt the user to select between the four types of register views, as shown in Figure L2–8. In both the memory and register views, a value can be modified by clicking on the required field and directly entering a new value.

46 C6x-Based DSP

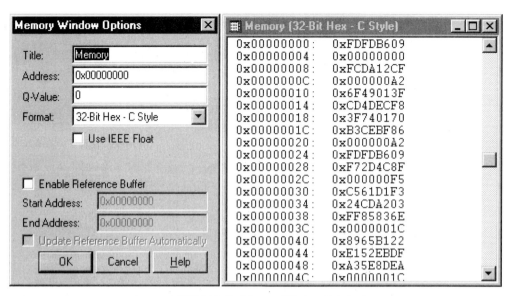

FIGURE L2–7 Memory Window dialog.

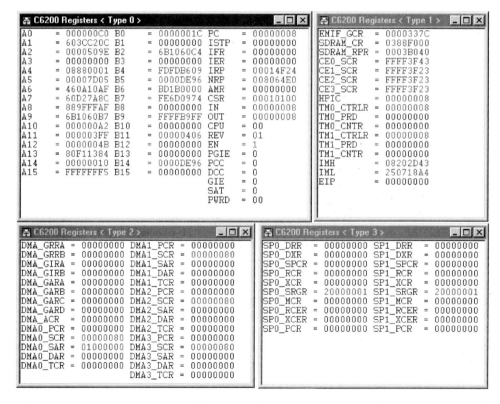

FIGURE L2–8 Register watch.

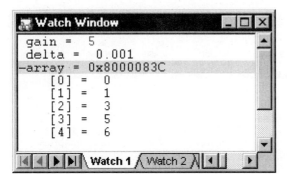

FIGURE L2–9 Watch Window.

To watch symbolic variables, the VIEW / WATCH WINDOW command is used. This command brings up the Watch Window. By right clicking in the window and selecting INSERT NEW EXPRESSION, symbolic variables can be added to the window. A Watch Window with two variables and an array is shown in Figure L2–9.

When a variable is out of scope, the stack option can be used to view the value of the variable. To view the stack, the command VIEW / CALL STACK is used. If a variable goes out of scope, the Watch Window will display "unknown identifier". Selecting the correct function from the stack window will allow the user to view the variable in the correct scope.

L2.3 Real-Time Analysis

By using DSP/BIOS Plug-ins and API functions, it is possible to monitor and analyze a program without disrupting its real-time behavior. The DSP / BIOS Plug-ins provide real-time analysis capabilities by program tracing, performance monitoring, and file streaming. Program tracing is made possible by writing to target logs using the API function LOG_printf. Performance monitoring tracks statistics such as processor load and thread timings. A thread is the term used to refer to events such as a hardware interrupt, a software interrupt, and an idle function or a periodic function. File streaming is used to bind target-resident I/O objects to host files. Figure L2–10 shows an execution chart, a CPU load graph, a message log window, and an external program using RTDX.

To include the DSP / BIOS capabilities in a project, the command FILE / NEW / DSP BIOS CONFIGURATION is used. This will create a new configuration file, as shown in Figure L2–11. The objects shown in the Configuration window are actually separate modules bound together by creating a single executable program. The result of a save from the configuration creates several files that must be included and linked for proper operation. When a configuration file is saved, a linker command file (.cmd), a header file (.h62), and an assembly source file (.s62) are created. The .cdb file must be included in the project. The command file will replace any other command file that is being used. The source file also eliminates the need to link to the *rts6201.lib* library.

The modules of the DSP /BIOS, as specified in [8], consist of the following:

- **CLK:** The on-chip timer module controls the on-chip timer and provides a logical 32-bit real-time clock with a high-resolution interrupt rate as short as the resolution of the on-chip timer register (four instruction cycles) and a low-resolution interrupt rate as long as several milliseconds.

48 C6x-Based DSP

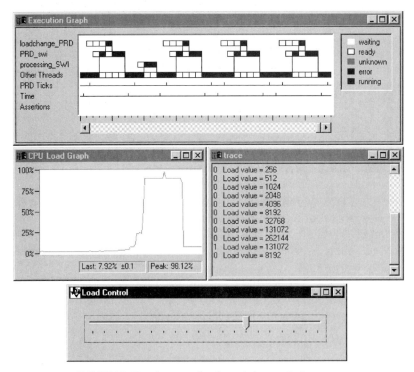

FIGURE L2–10   An example of a real-time analysis screen.

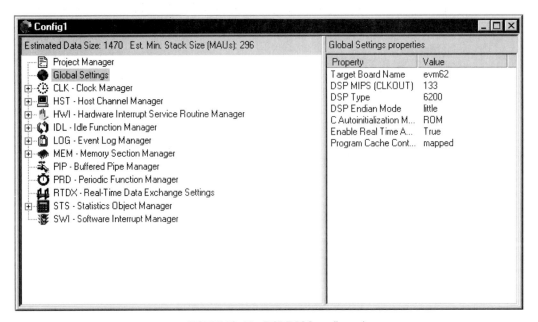

FIGURE L2–11   DSP/BIOS configuration.

- **HST:** The host input/output module manages host channel objects allowing an application to stream data between the target and the host. Host channels are statically configured for input or output.
- **HWI:** The hardware interrupt module provides support for hardware interrupt routines. In a configuration file, functions can be assigned to run when hardware interrupts occur.
- **IDL:** The idle function module manages idle functions, which run in a loop when the target program has no higher priority functions to perform.
- **LOG:** The log module manages LOG objects, capturing events in real-time while the target program executes. System logs or user-defined logs can be used, and messages in these logs can be viewed in real-time.
- **MEM:** The memory module allows one to specify the memory segments required to locate various code and data sections of a target program.
- **PIP:** The data pipe module manages data pipes, which are used to buffer streams of input and output data. These data pipes provide a consistent software data structure to drive I/O between the DSP and other devices.
- **PRD:** The periodic function module manages periodic objects, which trigger cyclic execution of program functions. The execution rate of these objects can be controlled by the clock rate maintained by the CLK module (usually in response to hardware interrupts from devices that produce or consume streams of data).
- **RTDX:** Real-time data exchange permits the data to be exchanged between the host and target in real-time and then be analyzed and displayed on the host using an OLE (object linking and embedding) automation client.
- **STS:** The software interrupt module manages software interrupts, which are patterned after hardware interrupts service routines. When a target program posts a SWI object with an API call, the SWI module schedules execution of the corresponding function. Software interrupts can have up to 15 priority levels.
- **TRC:** The trace module manages a set of trace control bits that control the real-time capture of program information through event and statistics accumulators.

To make the task of transferring data between the host and the target easy, the CCS uses RTDX through the JTAG (Joint Test Action Group) link. Using RTDX, it is possible to transfer data to and from the DSP without stopping its operation. This becomes very useful when data can be analyzed by an OLE automation client, allowing the user to speed up the design process.

CHAPTER 4

Sampling

Sampling constitutes the process of generating discrete time samples from an analog signal. First it is helpful to show the relationship between analog and digital frequencies. Let us consider the analog signal $x(t) = A\cos(\omega t + \phi)$. Sampling this signal at $t = nT_s$ generates the discrete time signal $x[n] = A\cos(\omega n T_s + \phi) = A\cos(\theta n + \phi)$, $n = 0, 1, 2, \ldots$, where $\theta = \omega T_s = \dfrac{2\pi f}{f_s}$ denotes digital frequency with units radians (as compared with analog frequency ω with units radians/sec). To see the difference between analog and digital frequencies, note that the same discrete time signal is obtained for different continuous-time signals if the product ωT_s remains the same as is shown in Figure 4–1. Likewise, different discrete time signals are obtained for the same analog or

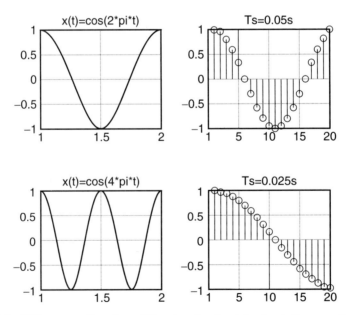

FIGURE 4–1 Different sampling of two different analog signals leading to the same digital signal.

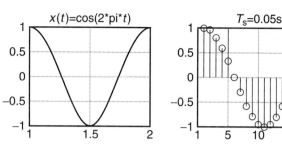
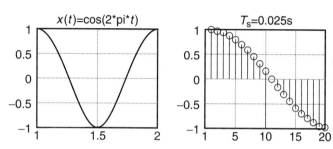

FIGURE 4–2 Different sampling of the same analog signal leading to two different digital signals.

continuous-time signal when the sampling rate is changed as shown in Figure 4–2. In other words, both the frequency of the analog signal and the sampling frequency define the frequency of the corresponding digital signal.

It helps to understand the constraints associated with the preceding sampling process by examining signals in frequency domain. As illustrated in Figure 4–3, when an analog signal with a maximum frequency of f_{max} is sampled at a rate of $T_s = \dfrac{1}{f_s}$, its

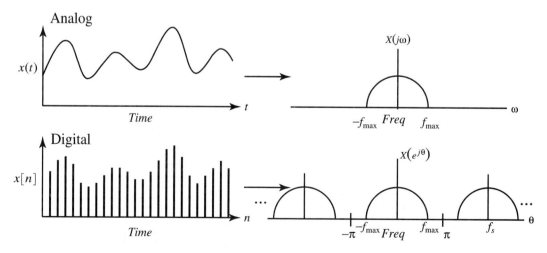

FIGURE 4–3 Signal sampling.

corresponding frequency response is repeated every 2π radians. In other words, the frequency response transform in discrete domain becomes a periodic version of the Fourier transform in analog domain. These transform pairs are now listed:

Fourier transform for nonperiodic analog signals

$$X(j\omega) = \int_{-\infty}^{\infty} x(t)e^{-j\omega t}\, dt,$$

$$x(t) = \frac{1}{2\pi}\int_{-\infty}^{\infty} X(j\omega)e^{j\omega t}\, d\omega.$$

Fourier series for periodic analog signals

$$X_k = \frac{1}{T}\int_{-T/2}^{T/2} x(t)e^{-j\omega_0 k t}\, dt,$$

$$x(t) = \sum_{k=-\infty}^{\infty} X_k e^{j\omega_0 k t},$$

where T denotes period and ω_0 fundamental frequency.

Fourier transform for nonperiodic digital signals

$$X(e^{j\theta}) = \sum_{n=-\infty}^{\infty} x[n]e^{-jn\theta}, \quad \theta = \omega T_s,$$

$$x[n] = \frac{1}{2\pi}\int_{-\pi}^{\pi} X(e^{j\theta})e^{jn\theta}\, d\theta.$$

Discrete Fourier tranform (DFT) for periodic digital signals

$$X[k] = \sum_{n=0}^{N-1} x[n]e^{-j\frac{2\pi}{N}nk}, \quad k = 0, 1, \ldots, N-1,$$

$$x[n] = \frac{1}{N}\sum_{k=0}^{N-1} X[k]e^{j\frac{2\pi}{N}nk}, \quad n = 0, 1, \ldots, N-1.$$

Therefore, to avoid any aliasing or distortion of the frequency content of the digital signal, and hence to be able to recover or reconstruct the frequency content of the original analog signal, we must have $f_s \geq 2f_{max}$. This is known as the Nyquist rate. That is, the sampling frequency should be at least twice the highest frequency in the signal. Normally, before any digital manipulation, an antialiasing lowpass analog filter is used to limit the highest frequency of the analog signal.

The aliasing problem can be further illustrated by considering an undersampled sinusoidal signal as shown in Figure 4-4. In this figure, a 1-KHz sinusoid is sampled at $f_s = 800$ samples/sec, which is less than the Nyquist rate. The dashed line curve is a 200-Hz sinusoid that passes through the same sample points. Thus, at this sampling frequency the output of the A/D converter would be the same if either of the sinusoids were the input signal.

4.1 SAMPLING ON THE C6X EVM

On the C6x EVM board, there is a 16-bit codec that performs the analog-to-digital conversion with sampling frequencies ranging from 5.5 to 48 KHz. The codec is fully pro-

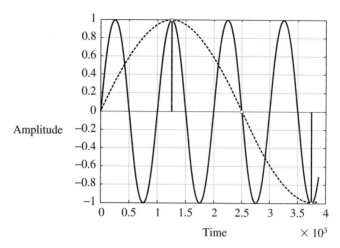

FIGURE 4–4 Ambiguity caused by aliasing.

grammable from C and assembly. However, it is easier to use the available C library functions to reset, initialize, and run the codec. There are numerous functions and macros that allow for the codec to be easily initialized and controlled.

Processing samples of a signal can be done within an ISR (interrupt service routine). Let us first explain what an interrupt is. As the name interrupt implies, the processor halts whatever it is processing and goes to a specific location in memory to execute an ISR. An interrupt can be issued externally or internally. There are four external interrupts and twelve internal interrupts possible on the C6x. These interrupts are prioritized. Note that RESET is considered to be an interrupt having the highest priority. Refer to the TI TMS320C6x CPU and Instruction Set Reference Guide [1] for interrupt priorities.

The location the processor will go to after an interrupt occurs corresponds to a predefined offset for that interrupt added to the Interrupt Service Table (ISTB) as part of the Interrupt Service Table Pointer (ISTP) register. As an example, for the receive interrupt RINT0 (used to process samples in the lab exercises), the processor would go to the location ISTB+0x1A0. At this location, there is normally a branch instruction that would take the processor to a receive ISR somewhere in memory, as is shown in Figure 4–5.

In general, an ISR includes three parts. The first and last part incorporate saving and restoring registers, respectively. The actual interrupt routine makes up the second part. If needed, saving and restoring are done to restore the status of the processor to the time when the interrupt was issued.

Interrupts can be enabled or disabled by setting or clearing appropriate bits in the Interrupt Enable Register (IER). There is a master switch, Global Interrupt Enable bit (GIE) as part of the Control Status Register (CSR), which can be used to turn all inter-

54 C6x-Based DSP

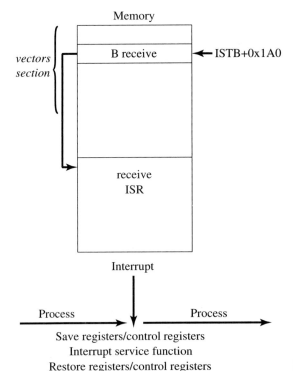

FIGURE 4–5 Interrupt service routine.

rupts on or off. For example, the assembly code shown in Figure 4–6 indicates how to enable INT4, the GIE bit. Here the instruction MVC (move to and from a control register) is used to transfer a control register to a CPU register for bit manipulation. Another register called the Interrupt Flag Register (IFR) allows one to check if or what interrupt has occurred. (See the CPU manual [1] for more details on interrupt control registers.)

LAB 3: AUDIO SIGNAL SAMPLING

The purpose of this lab is to use the C62 evaluation module board to sample an analog audio signal in real-time. A common approach to handling real-time signals, which is the use of an interrupt service routine, is utilized here to obtain and process signal samples.

```
MVK    .S2    0010h, B3      ; bit4="1"
MVC    .S2    IER, B4        ; get IER
OR     .L2    B3, B4, B4     ; set bit4
MVC    .S2    B4, IER        ; write IER
MVC    .S2    CSR, B5        ; get CSR_GIE
OR     .L2    1, B5, B5      ; bit0="1"
MVC    .S2    B5, CSR        ; set GIE
```

FIGURE 4–6 Setup code to recognize INT4.

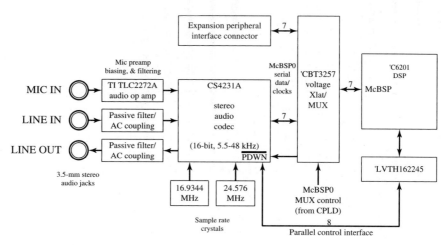

FIGURE L3–1 EVM stereo audio interface.[†]

The C62 EVM has a 16-bit stereo audio coder/decoder CS4231A (codec) that can handle sampling rates from 5.5 KHz to 48 KHz. There are three 3.5-mm audio jacks on the back of the EVM that allow for a microphone in, a line in, and a line out. The codec is connected to the C62 DSP through its multi-channel buffered serial port (McBSP). Each of the audio jacks has its own amplifying and filtering capabilities. A block diagram of the EVM stereo interface is shown in Figure L3–1.

All the adjustments to the codec can be done through the DSP support software provided by Texas Instruments. The DSP support software contains C functions that are used to work with the McBSP, codec, and various EVM board utilities. The codec library is supplied in the archived object library file *drv6x.lib* or *drv6xe.lib,* which is the big-endian version of the library. The source for the library is contained in the file *drv6x.src*. The codec library has so-called API (application program interface) functions that can be used to configure and control the operation of the codec. The functional descriptions of these functions can be found in the EVM User's Guide [2] under TMS320C6x EVM DSP Support Software. These functions are utilized here to write a sampling program for the C62 EVM.

L3.1 Initializing EVM and Codec

In writing a program that uses the codec to sample an incoming analog signal, several initializations have to be performed. Among these are the initialization of the EVM, McBSP, and codec. The sampling rate for the codec has to be set, as well as any gain adjustments in the input ADC and output DAC (digital-to-analog converter). The API functions can be used to achieve all of these mentioned adjustments. Once the required initializations are made, an interrupt needs to be assigned to the receive register of the codec to halt the processor and jump to an assigned interrupt service routine. The final program will output the same input sample to the codec for audio or video display. The following program includes an order of API functions that achieves all of the foregoing mentioned initializations:

56   C6x-Based DSP

```c
#include <stdlib.h>
#include <stdio.h>
#include <string.h>
#include <common.h>
#include <mcbspdrv.h>
#include <intr.h>
#include <board.h>
#include <codec.h>
#include <mcbsp.h>
#include <mathf.h>

void hookint(void);
interrupt void serialPortRcvISR (void);

void main()
{
        Mcbsp_dev dev;
        Mcbsp_config  mcbspConfig;
        int sampleRate,actualRate;

        /**************************************************************/
        /* Initialize EVM                                             */
        /**************************************************************/

        evm_init();

        /**************************************************************/
        /* Open MCBSP                                                 */
        /**************************************************************/

        mcbsp_drv_init();

        dev = mcbsp_open(0);

        if (dev == NULL)
        {
                printf("Error opening MCBSP 0    \n ");
                return(ERROR);
        }

        /**************************************************************/
        /* configure McBSP                                            */
        /**************************************************************/

        memset(&mcbspConfig,0,sizeof(mcbspConfig));

        mcbspConfig.loopback                 = FALSE;

        mcbspConfig.tx.update                = TRUE;
```

Chapter 4   Sampling

```c
    mcbspConfig.tx.clock_mode           = CLK_MODE_EXT;
    mcbspConfig.tx.frame_length1        = 0;
    mcbspConfig.tx.word_length1         = WORD_LENGTH_32;

    mcbspConfig.rx.update               = TRUE;
    mcbspConfig.rx.clock_mode           = CLK_MODE_EXT;
    mcbspConfig.rx.frame_length1        = 0;
    mcbspConfig.rx.word_length1         = WORD_LENGTH_32;

    mcbsp_config(dev,&mcbspConfig);

    MCBSP_ENABLE(0, MCBSP_BOTH);

    /****************************************************************/
    /* configure CODEC                                              */
    /****************************************************************/
    codec_init();

    /* A/D 0.0 dB gain, turn off 20dB mic gain, sel (L/R)LINE input*/
    codec_adc_control(LEFT,0.0,FALSE,LINE_SEL);
    codec_adc_control(RIGHT,0.0,FALSE,LINE_SEL);

    /* mute (L/R)LINE input to mixer                   */
    codec_line_in_control(LEFT,MIN_AUX_LINE_GAIN,TRUE);
    codec_line_in_control(RIGHT,MIN_AUX_LINE_GAIN,TRUE);

    /* D/A 0.0 dB atten, do not mute DAC outputs            */
    codec_dac_control(LEFT, 0.0, FALSE);
    codec_dac_control(RIGHT, 0.0, FALSE);

    sampleRate = 44100;

    actualrate = codec_change_sample_rate(sampleRate, TRUE);

    codec_interrupt_enable();

    hookint();




    /****************************************************************/
    /* Main Loop, wait for Interrupt                                */
    /****************************************************************/
    while (1)
    {
    }
}
```

```
void hookint()
{
        intr_init();
        intr_map(CPU_INT15, ISN_RINT0);
        intr_hook(serialPortRcvISR, CPU_INT15);

        INTR_ENABLE(15);
        INTR_GLOBAL_ENABLE();
        return;
}

interrupt void serialPortRcvISR (void)
{
        int temp;

        temp = MCBSP_READ(0);
        MCBSP_WRITE(0, temp);
}
```

This program is explained here in a step-by-step fashion. We must first initialize the EVM. This is done by using the function evm_init() before calling any other support functions. This function configures the EVM base address variables, and initializes the external memory interface (EMIF). The return value of this function indicates success or failure for the initialization of the EVM.

Once the EVM has been successfully initialized, the next step is to open a handle to the McBSP in order to send and receive data. The McBSP API functions are used for this purpose. The function mcbsp_drv_init() initializes the McBSP driver and allocates memory for the device handle. The return value of this function also indicates success or failure. After the initialization of the McBSP driver, the data structure elements that control the behavior of the McBSP are set to their default values. (For more details, refer to the EVM User's Guide [2].) Then, the McBSP needs to be actually opened to get a handle to it. The function mcbsp_open() is used to return the handle *dev* in order to control the McBSP.

The next item would be to adjust the parameters of the McBSP. The data structure of the McBSP gets initialized to its default values as a result of using the initialization functions, so all we have to do is adjust several parameters to suit our needs. The loopback property of the codec has to be turned off, and the data and frame lengths have to be set. The mode of the clock has to be specified as well. The adjustments to the codec are made by allocating memory to the structure Mcbsp_Config using the memset() function and then setting the required variables. The address of this structure is then passed as an argument to the mcbsp_config() function, which performs the required adjustments. The following lines of code are used to do these adjustments. Finally, the McBSP has to be activated. This is done by using the macro MCBSP_ENABLE(), which is defined in the *mcbsp.h* header file. A macro is a collection of instructions that gets substituted in the code by the assembler.

```
/************************************************************/
/* configure McBSP                                          */
/************************************************************/
memset(&mcbspConfig,0,sizeof(mcbspConfig));

mcbspConfig.loopback            = FALSE;

mcbspConfig.tx.update           = TRUE;
mcbspConfig.tx.clock_mode       = CLK_MODE_EXT;
mcbspConfig.tx.frame_length1    = 0;
mcbspConfig.tx.word_length1     = WORD_LENGTH_32;

mcbspConfig.rx.update           = TRUE;
mcbspConfig.rx.clock_mode       = CLK_MODE_EXT;
mcbspConfig.rx.frame_length1    = 0;
mcbspConfig.rx.word_length1     = WORD_LENGTH_32;

mcbsp_config(dev,&mcbspConfig);

MCBSP_ENABLE(0, MCBSP_BOTH);
```

The codec can now be initialized by using the `codec_init()` function from the codec API. This function sets default parameters to the codec and calibrates it. The next step is to adjust the parameters of the codec. The main parameter to adjust here is sampling rate. This is done by using the `codec_change_sample_rate()` function. This function sets the sampling rate of the codec to the closest allowed sampling rate of the passed argument. The return value from this function is the actual sampling rate. The other required adjustments are the selection of the line-in or mic-in ports and the adjustment of their gain settings. To have stereo input, both channels are selected and their gains are adjusted to 0 dB settings. The functions that accomplish these tasks are `codec_adc_control()`, `codec_line_in_control()`, and `codec_dac_control()`. The following lines of code are used for the purpose of initializing the codec. It is also required for the codec to generate interrupts as data is received in the DRR (data receive register). Hence, the interrupt processing capability of the codec must be enabled. This is accomplished by using the `codec_interrupt_enable()` function.

```
/************************************************************/
/* configure CODEC                                          */
/************************************************************/
codec_init();
// ADC 0.0 dB gain, turn off 20dB mic gain, sel (L/R)LINE input
codec_adc_control(LEFT,0.0,FALSE,LINE_SEL);
codec_adc_control(RIGHT,0.0,FALSE,LINE_SEL);
// (L/R) LINE input to mixer
codec_line_in_control(LEFT,MIN_AUX_LINE_GAIN,FALSE);
codec_line_in_control(RIGHT,MIN_AUX_LINE_GAIN,FALSE);
```

```
            // DAC 0.0 dB atten, do not mute DAC outputs
            codec_dac_control(LEFT, 0.0, FALSE);
            codec_dac_control(RIGHT, 0.0, FALSE);

            sampleRate = 44100;
            actualrate = codec_change_sample_rate(sampleRate, TRUE);

            codec_interrupt_enable();
```

The initialization of the EVM, McBSP, and codec is now complete. Next, let us focus our attention on setting up an interrupt that will branch to a simple ISR to handle an incoming sampled signal.

L3.2 Interrupt Data Processing

The idea of using interrupts to halt the DSP is commonly used in real-time data processing. This approach is widely used, since it eliminates the need for complicated synchronization schemes. In our case, the processor is halted as it resides in an infinite wait state to indicate that a new data sample has arrived and is waiting in the DRR of the serial port. The generated interrupt will branch to an ISR, which can then be used to process the sample and send it back out. To do this, the interrupt capabilities of the EVM must be enabled and adjusted so that an unused interrupt is assigned to the DRR event of the serial port. The interrupt is then set (hooked) to the ISR.

The first task at hand is to initialize the interrupt service table pointer (ISTP) with the address of the global `vec_table` to be resolved at the link time. This is done by placing the base address of the vector table in the ISTP. The `intr_init()` function is used for this purpose. Next, we need to select an interrupt number and map it to a CPU interrupt, which, in our case, is the DRR receive interrupt. Here, the CPU interrupt 15 is used and mapped to the DRR interrupt by using the `intr_map()` function. To connect an ISR to the interrupt that we have set up, the `intr_hook()` function is used, and the name of the function that we wish to use is passed to it. The last thing is to enable the interrupts by using the macros INTR_ENABLE and INTR_GLOBAL_ENABLE. The following lines of code initializes the CPU interrupt 15 to the DRR interrupt and hooks it to an ISR named *serialPortRcvISR*.

```
            intr_init();
            intr_map(CPU_INT15, ISN_RINT0);
            intr_hook(serialPortRcvISR, CPU_INT15);

            INTR_ENABLE(15);
            INTR_GLOBAL_ENABLE();
```

A simple ISR can now be used to receive samples from the McBSP and send them back out, unprocessed for the time being. To define an ISR, we need to use the interrupt declaration at the beginning of the function. The macros MCBSP_READ and MCBSP_WRITE are used to read samples from the DRR and write them to the DXR (Data Transmit Register) of the McBSP. Such an ISR is now presented:

```
interrupt void serialPortRcvISR (void)
{
        int temp;

        temp = MCBSP_READ(0);

        MCBSP_WRITE(0, temp);
}
```

Since the CPU is not actually doing anything as it waits for a new data sample, we can have an infinite loop inside the main program to keep it running. As an interrupt comes in from the DRR register, the program branches to the ISR, performs it, and then returns to its wait state. This is accomplished by using a simple while(1){ } statement.

Now the complete code for a sampling program that samples an analog signal, such as the output from a CD player connected to the line-in port of the EVM and generates an interrupt that calls a simple ISR, is ready for use.

The program can be built by using the following commands:

Build commands
```
cl6x -c -ealst -ss -op0 -g -i. -iC:\Evm6x\Dsp\Include -
iC:\C6xTools\include   codec.c

lnk6x -m codec.map -heap 0x00300000 -c -o codec.out codec.obj
   C:\Evm6x\Dsp\Lib\Drivers\Drv6x.lib
   C:\Evm6x\Dsp\Lib\DevLib\Dev6x.lib
   C:\C6xTools\lib\rts6201.lib
   link.cmd
```

link.cmd
```
MEMORY
{
 INT_PROG_MEM    (RX)    : origin = 0x00000000 length = 0x00010000
 SBSRAM_PROG_MEM (RX)    : origin = 0x00400000 length = 0x00014000
 SBSRAM_DATA_MEM (RW)    : origin = 0x00414000 length = 0x0002C000
 SDRAM0_DATA_MEM (RW)    : origin = 0x02000000 length = 0x00400000
 SDRAM1_DATA_MEM (RW)    : origin = 0x03000000 length = 0x00400000
 INT_DATA_MEM    (RW)    : origin = 0x80000000 length = 0x00010000
}
```

```
SECTIONS
{
    .vec:        load = 0x00000000
    .text:       load = SBSRAM_PROG_MEM
    .const:      load = INT_DATA_MEM
    .bss:        load = INT_DATA_MEM
    .data:       load = INT_DATA_MEM
    .cinit       load = INT_DATA_MEM
    .pinit       load = INT_DATA_MEM
    .stack       load = INT_DATA_MEM
    .far         load = INT_DATA_MEM
    .sysmem      load = SDRAM0_DATA_MEM
    .cio         load = INT_DATA_MEM
    sbsbuf       load = SBSRAM_DATA_MEM
           { _SbsramDataAddr = .; _SbsramDataSize = 0x0002C000; }
}
```

By running the EVM debugger and connecting the output of a CD player to the line-in port and a pair of powered speakers to the line-out port, CD-quality sound should be heard. It may be required to change some of the paths to access the required header and library files.

CHAPTER 5

Fixed-Point vs. Floating-Point

One important feature that distinguishes different DSPs is whether their CPUs perform fixed-point or floating-point arithmetic. In a fixed-point processor, numbers are represented and manipulated in 2's-complement integer format. In a floating-point processor, in addition to integer arithmetic, floating-point arithmetic can be handled. This means that numbers are represented by the combination of a mantissa, or a fractional part, and an exponent part, and the CPU possesses the necessary hardware for manipulating both of these parts. As a result, in general, floating-point processors are more expensive and slower than fixed-point ones.

In a fixed-point processor, one needs to be concerned with the dynamic range of numbers, since a much narrower range of numbers can be represented in the integer format than the floating-point format. For most applications, such a concern can be virtually ignored when using a floating-point processor. Consequently, fixed-point processors usually demand more coding effort than do floating-point processors.

5.1 Q-FORMAT NUMBER REPRESENTATION ON FIXED-POINT PROCESSORS

The decimal value of a 2's-complement number $B = b_{N-1}b_{N-2}\ldots b_1 b_0$, $b_i \in \{0, 1\}$, is given by

$$D(B) = -b_{N-1}2^{N-1} + b_{N-2}2^{N-2} + \ldots + b_1 2^1 + b_0 2^0.$$

The 2's-complement representation allows the processor to perform addition and subtraction using the same hardware. In many algorithms, numbers need to be normalized between -1 and 1 or as fractions in order to cope with the limitation imposed by the representation just given. Considering that products of fractions would still be a fraction, this normalization is achieved by the programmer moving the implied or imaginary binary point, as shown in Figure 5–1. (Note that there is no physical memory allocated to this point.) Hence, in this way, the fractional value is given by

$$F(B) = -b_{N-1}2^0 + b_{N-2}2^{-1} + \ldots + b_1 2^{-(N-2)} + b_0 2^{-(N-1)}.$$

This representation scheme is referred to as Q-format, or fractional representation. The programmer needs to keep track of the implied binary point when manipulating Q-format numbers. For instance, let us consider two Q-15 format numbers when using

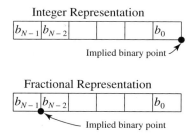

FIGURE 5–1 Number representations.

a 16-bit wide memory. Each number consists of 1 sign bit plus 15 fractional bits. When these numbers are multiplied, a Q-30 format number is resulted (the product of two fractions is still a fraction) with bit 31 being the sign bit and bit 32 another sign bit (called extended sign bit). If not enough bits are available to store all 32 bits and only 16 bits can be stored, it makes sense to store the most significant bits. This translates into storing the upper portion of the 32-bit product register by doing a 15-bit right shift (SHR). In this manner, the product would be stored in Q-15 format. (See Figure 5–2.)

Using the 2's-complement representation, a dynamic range of $-(2^{N-1})$ $\leq D(B) \leq 2^{N-1} - 1$ can be achieved, where N denotes the number of bits. As an example, let us consider the four-bit case where the most negative number is -8 and the most positive number is 7. The decimal representations of the numbers are shown in Figure 5–3. Notice how the numbers change from most positive to most negative with the sign bit.

For the four-bit case, we are limited by -8 and 7, so it is easy to see that any multiplication or addition resulting in a number larger than 7 or smaller than -8 will cause

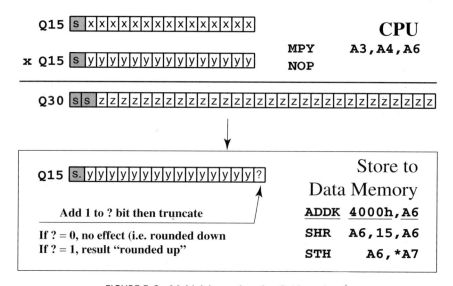

FIGURE 5–2 Multiplying and storing Q-15 numbers.[†]

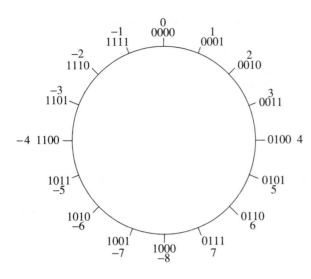

FIGURE 5–3 Four-bit binary representation.

overflow. For example, when 6 is multiplied by 2, we get 12. Hence, the result is greater than the representation limits and will be "wrapped" around the circle to 1100, which is −4.

The Q-format representation solves this problem by normalizing the dynamic range between −1 and 1. Any resulting multiplication will be within the limits of the dynamic range. Using the Q-format representation, the dynamic range is divided into 2^N sections, where $2^{-(N-1)}$ is the size of a section. The most negative number is always −1, and the most positive number is $1 - 2^{-(N-1)}$.

The next example helps one to see the difference in the two representations. As shown in Figure 5–4, the multiplication of 1110 by 0110 in binary is the equivalent of multiplying −2 by 6 in decimal, which results in −12, a number exceeding the dynamic range of the four-bit system. Based on the Q-3 representation, these numbers correspond to −0.25 and 0.75. The result is −0.1875, which falls within the fractional range.

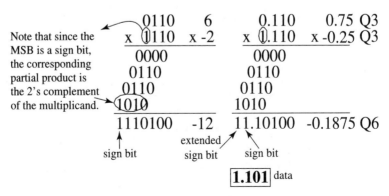

FIGURE 5–4 Binary and fractional multiplication.

Notice that the hardware generates the same 1's and 0's; what is different is the interpretation of the bits.

When multiplying Q-N numbers, it should be remembered that the result will be $2(N + 1)$ bits in length. According to the amount of storage space available, the result has to be shifted accordingly. If two Q-15 numbers are multiplied, the result will be 32 bits with the MSB (most significant bit) being the extended sign bit followed by the sign bit. The imaginary decimal point will be after the 30th bit. So, a right shift of 15 is required to store it in a 16 bit memory. If a 32 bit storage capability is available, a left shift of one can be done to remove the extended sign bit.

Although Q-format solves the problem of overflow in multiplication, addition and subtraction still pose a problem. When adding, say, 0.5 and 0.75, the result is 1.25, which exceeds the range of any Q-format representation. To solve this problem, the scaling approach discussed in Lab 4 can be employed.

5.2 FINITE WORD LENGTH EFFECTS ON FIXED-POINT PROCESSORS

Due to the fact that memory or registers have finite number of bits, there could be a noticeable error between desired and actual outcomes on a fixed-point processor. The so-called finite word length quantization effect is similar to input data quantization effect introduced by an A/D converter. It helps to understand the finite word length effect by first discussing the process of quantization as introduced by an A/D converter possessing a finite resolution (number of bits).

5.2.1 Input or A/D Quantization

The quantization interval depends on the number of quantization or resolution level, as shown in Figure 5–5. Based on a full-scale voltage range of $2V_{ref}$ and a $b + 1$ bit quantizer, the quantization interval δ is given by (assuming uniform quantization)

$$\delta = \frac{2V_{ref}}{2^{b+1}}.$$

Clearly the amount of quantization noise generated by the A/D conversion depends on the size of the quantization interval. The presence of more quantization bits translates into having a narrower quantization interval, and hence into a lower amount of quantization

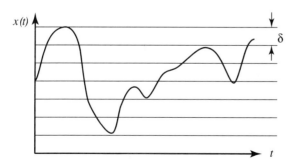

FIGURE 5–5 Quantization levels.

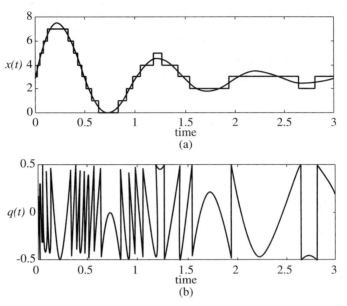

FIGURE 5-6 (a) Continuous to discrete voltage conversion (b) corresponding quantization noise signal.

noise. Figure 5–6 illustrates a typical quantization noise signal caused by the A/D conversion.

Assuming the quantization noise signal $q(t)$ is uniformly distributed between $-\frac{\delta}{2}$ and $\frac{\delta}{2}$, the corresponding quantization noise variance or power is given by $\sigma_q^2 = E[q^2(t)] = \frac{\delta^2}{12}$. The signal-to-noise ratio in dB as introduced by the A/D conversion can be written as $\text{SNR}_{A/D} = 10 \log_{10} \frac{\sigma_x^2}{\sigma_q^2}$, where σ_x^2 denotes the signal variance.

To better understand the quantization effect, let us assume that the signal is zero-mean Gaussian with $V_{\text{ref}} = K\sigma_x$. By substituting for V_{ref} and σ_x in the preceding equation, we obtain $\text{SNR}_{A/D}(\text{dB}) \cong 6b + 16.8 - 20 \log_{10} K$. This means that increasing the quantization by one bit raises the signal-to-quantization-noise ratio by 6 dB. As an example, consider $K = 4$. The probability of the signal samples falling in the $4\sigma_x$ range is 0.954. This means that out of 1,000 samples, 954 samples fall in this range on the average. In other words, 46 out of 1,000 samples fall outside the indicated range and hence get represented by the maximum or minimum allowable value. So when using a 16-bit converter (or $b = 15$), the signal-to-noise ratio as a consequence of the A/D conversion is 95 dB.

If the signal is scaled by α, the corresponding signal variance changes to $\alpha^2 \sigma_x^2$. Hence, the foregoing signal-to-quantization-noise ratio changes to

$$\text{SNR}_{A/D} \cong 6b + 16.8 - 20 \log_{10} K + 20 \log_{10} \alpha.$$

It is important to note that when we perform fractional arithmetic, α is scaled to be less than 1, which leads to a lower signal-to-noise ratio. This indicates that the quantization noise should be kept in mind when scaling down the input signal. In other words, scaling down to achieve fractional representation cannot be done indefinitely, since, as a result, the signal would get buried in the quantization noise. The interested reader is referred to [12] for more elaborate analysis of this noise source. For example, for a linear time-invariant system such as a FIR or an IIR filter, it can be shown that the noise variance σ_o^2 at the output of the system as caused by the input quantization noise is given by [12]

$$\sigma_o^2 = \sigma_q^2 \sum_k h^2[k],$$

where h denotes the unit sample response. For a first-order system, $y[n] = Ay[n-1] + x[n]$, it is easy to show that

$$\sigma_o^2 = \frac{\sigma_q^2}{1 - A^2}.$$

In other words, as the pole gets closer to the unit circle, the quantization noise effect at the output is enhanced.

5.2.2 Finite Word Length Error

Consider fractional numbers quantized by a $b + 1$ bit converter. When these numbers are manipulated and stored in a $M + 1$ bit memory, $M < b$, there is going to be an error (due to the fact that $b - M$ of the least significant fractional bits are discarded or truncated). This so-called finite-word-length error could unacceptably alter the behavior of a system. The range of the magnitude of this truncation error ξ_t is given by $0 \leq |\xi_t| \leq 2^M - 2^b$. The lowest level of truncation error corresponds to the situation when all the thrown-away bits are zeros, while the highest level of truncation error corresponds to the situation when all the thrown away bits are ones. This effect has been extensively studied for FIR and IIR filters. (For example, see [13].) Since the coefficients of such filters are represented by a finite number of bits, the roots of transfer function polynomials or the positions of zeros and poles shift in the complex plane. The amount of shift in the positions of poles and zeros can be related to the amount of quantization error in the coefficients. For example, for an Nth-order IIR filter, the sensitivity of the ith pole p_i with respect to the kth coefficient A_k can be derived to be (see [14]),

$$\frac{\partial p_i}{\partial A_k} = \frac{-p_i^{N-k}}{\prod_{\substack{l=1 \\ l \neq i}}^{N} (p_i - p_l)}.$$

This means that the change in the position of a pole is influenced by the positions of all other poles. That is the reason the implementation of an Nth-order IIR filter is nor-

FIGURE 5–7 Floating-point data representation.

mally achieved by having a number of second-order IIR filters in series in order to decouple this dependency of poles. Also, note that as a result of coefficient quantization, the actual frequency response $\hat{H}(e^{j\theta})$ would become different than the desired frequency response $H(e^{j\theta})$. For example, for a FIR filter having N coefficients, it can be easily shown that the amount of error in the magnitude of the frequency response, $|\Delta H(e^{j\theta})|$, is bounded by

$$|\Delta H(e^{j\theta})| = |H(e^{j\theta}) - \hat{H}(e^{j\theta})| \leq N2^{-b}.$$

In addition to the effects just mentioned, coefficient quantization can lead to limit cycles. This means that in the absence of an input, the response of a supposedly stable system (poles inside the unit circle) to a unit sample is oscillatory instead of diminishing in magnitude.

5.3 FLOATING-POINT NUMBER REPRESENTATION

Due to relatively limited dynamic ranges of fixed-point processors, when using such processors one should be concerned with the scaling issue or how big the numbers get in the manipulation of a signal. Scaling is not an issue when using floating-point processors, since the floating-point hardware allows a much wider dynamic range. The C67x processor is the floating-point version of the C6x family, with many additional floating-point instructions. There are two floating-point data representations on the C67x processor: single-precision (SP) and double precision (DP). In the single-precision format, a value is expressed as

$$-1^s \times 2^{(\exp - 127)} \times 1.\text{frac},$$

where s denotes sign bit (bit 31), exp exponent bits (bits 23 through 30), and frac fractional, or mantissa, bits (bits 0 through 22). (See Figure 5–7.)

Consequently, numbers as big as 3.4×10^{38} and as small as 1.175×10^{-38} can be processed. In the double-precision format, more fractional and exponent bits are made available by using two words as

$$-1^s \times 2^{(\exp - 1023)} \times 1.\text{frac},$$

where exponent bits are from bit 20 through 30 and fractional bits are all the bits of one word and bit 0 through 19 of the other word. (See Figure 5–8.) In this manner, numbers as big as 1.7×10^{308} and as small as 2.2×10^{-308} can be handled.

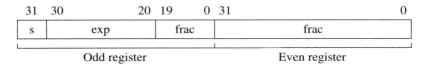

FIGURE 5–8 Double-precision floating-point representation.

.S Unit	
ABSSP	CMPLTDP
ABSDP	RCPSP
CMPGTSP	RCPDP
CMPEQSP	RSQRSP
CMPLTSP	RSQRDP
CMPGTDP	SPDP
CMPEQDP	

.L Unit	
ADDDP	INTSP
ADDSP	INTSPU
DPINT	SPINT
DPSP	SPTRUNC
INTDP	SUBSP
INTDPU	SUBDP

.M Unit	
MPYSP	MPYI
MPYDP	MPYID

ADDAD	LDDW

Note: Refer to the 'C62x/C67x CPU Reference Guide for more details

FIGURE 5-9 C67x additional floating-point instructions.[†]

When using a floating-point processor, all the steps needed to perform floating-point arithmetic are done by the CPU floating-point hardware. For example, consider adding two floating-point numbers represented by

$$a = a_{\text{frac}} \times 2^{a_{\text{exp}}},$$
$$b = b_{\text{frac}} \times 2^{b_{\text{exp}}}.$$

The floating-point sum c has the following exponent and fractional parts:

$$c = a + b,$$
$$= \left(a_{\text{frac}} + \left(b_{\text{frac}} \times 2^{-(a_{\text{exp}}-b_{\text{exp}})}\right)\right) \times 2^{a_{\text{exp}}}, \quad \text{if } a_{\text{exp}} \geq b_{\text{exp}},$$
$$= \left(\left(a_{\text{frac}} \times 2^{-(b_{\text{exp}}-a_{\text{exp}})}\right) + b_{\text{frac}}\right) \times 2^{b_{\text{exp}}}, \quad \text{if } a_{\text{exp}} < b_{\text{exp}}.$$

These parts are computed by the floating-point hardware. This shows that, though possible, it is inefficient to perform floating-point arithmetic on fixed-point processors. The instructions ending in SP denote single-precision data format and in DP denote double-precision data format (e.g., MPYSP and MPYDP). It should be noted that some of these instructions require additional execute (E) cycles, or latencies, as compared with fixed-point instructions. (Refer to Figure 2-9.) For example, MPYSP requires three delays or NOPs and MPYDP nine delays or NOPs as compared with one delay or NOP for fixed-point multiplication MPY. Figure 5-9 provides a listing of the floating-point instructions.

LAB 4: Q-FORMAT AND OVERFLOW

Fixed-point processors have a much smaller dynamic range than their floating-point counterparts. Even though C62 is considered to be a 32-bit device, its multiplier is only 16 bit. It is due to this limitation that a Q-15 representation of numbers is normally considered. The 16-bit multiplier can multiply two Q-15 numbers and produce a 32-bit result. Then, from there, the result can be stored in 32 bits or shifted back to 16 bits for storage or processing.

When multiplying two Q-15 numbers, which are in the range of −1 and 1, it is clear that the resulting number will always be in the same range. However, when two Q-15 numbers are added, the sum may fall outside this range, leading to overflow. Overflow can cause major problems by generating erroneous results. When using a fixed-point processor, the range of numbers must be closely examined and adjusted to compensate for overflow. The simplest correction method for overflow is to scale the input. This lab demonstrates the problem of overflow and shows the use of a scaling algorithm to correct for overflow.

L4.1 Scaling

The idea of scaling may be applied to most filtering and transform operations where the input is scaled down to do processing and the result is then scaled back up to the original size. An easy way to do this on the C62 is to use shifting. Since a right shift of 1 is equivalent to division by 2, we can scale the input by 2 repeatedly until all overflows disappear. The output can then be rescaled back to the total scaling amount.

It is often difficult to detect where in a program an overflow occurs. The C62 does not have an overflow detection hardware. It does, however, have a saturation flag. When using a saturated add instruction (SADD), overflow causes the saturation bit to be set to 1. To use this bit, hand-coded assembly is required where all the additions must be changed to SADD and monitored by reading the CSR. This is normally difficult, since most algorithms are directly coded in C.

Consider a simple multiply/accumulate operation. Assume that there are four constants that have to be multiplied by the input from the codec, which may be any analog signal. The worst possible case for overflow would be the case where all the multiplicants (C_k's and $x[n]$'s) are 1. For this case, the result will be 4, and $y[n] = \sum_{k=1}^{4} C_k * x[n-k]$.

Assuming that we only have control over the input $x[n]$, we have to scale the input so that the result $y[n]$ will fall in the dynamic range. A single right shift reduces the input to 0.5, and a double shift reduces it further to 0.25. Obviously, if we cannot control the C_k's, we can solve the overflow problem by further scaling down the input. Of course, this leads to less precision, but it is better than getting an erroneous result.

A simple method to implement this idea on the C62 would be to create a function that returns the necessary amount of scaling on the input. For any multiply/accumulate operation, such as filtering or transforms, the worst case is the multiplication and addition of all ones. Then, the required amount of scaling would be dependent on the number of additions in the summation. If we know the constants in the summation, such as filter coefficients or FFT twiddle factors, we can achieve a more precise result, since we would not have to do the maximum shift for all values.

To perform such a detection, a new addition function that is able to detect overflow is required. Given that all the inputs are in Q-15 format, the following piece of code achieves this detection by checking the CSR saturation bit after using the _sadd intrinsic:

```
short safe_add(short A, short B,int *status)
{
        int X,Y,result, SAT_BIT;
        X = A << 16;
        Y = B << 16;
        result = _sadd(X,Y);
        SAT_BIT=GET_REG_BIT(CSR,9);
        if(SAT_BIT==1){
                //Overflow Occured
                RESET_REG_BIT(CSR,9);          //Reset Sat Bit
                *status = 1;
        }
        else
                *status = 0;

        return (result >> 16);
}
```

This function adds two 16-bit numbers and reports any occurring overflow. If an overflow occurs, it also clears the SAT bit in the CSR. To create a simple scaling program, all that is needed is to use the preceding function to examine the worst-case scenario on a test array. The test array should be the one used in the algorithm, such as the coefficients of a filter. The function returns a single value indicating the number of times the input has to be scaled down to avoid any overflow.

Consider the constants shown in Table L4–1. Suppose the constants C_k's are to be used in a multiply/accumulate operation. Notice that in the worst-case scenario, the inputs are all ones, so the sum of C_k's overflows at the fourth summation, which is highlighted in the table. However, if the inputs are scaled down by one half, equivalent to scaling down the constants, the overflow disappears. The program in Figure L4–1 illustrates this scaling process.

TABLE L4–1 Scaling example.

C_k	$\sum C_k$	$\dfrac{C_k}{2}$	$\sum \dfrac{C_k}{2}$
0	0	0	0
0.8311	0.8311	0.41555	0.41555
−0.2977	0.5334	−0.14885	0.2667
0.4961	**1.0295**	0.24805	0.51475
0.6488	**1.6783**	0.3244	0.83915
−0.3401	**1.3382**	−0.17005	0.6691
−0.0341	**1.3041**	−0.01705	0.65205
−0.2336	**1.0705**	−0.1168	0.53525
−0.3801	0.6904	−0.19005	0.3452
−0.3984	0.292	−0.1992	0.146
−0.2568	0.0352	−0.1284	0.0176
0.4884	0.5236	0.2442	0.2618
−0.1113	0.4123	−0.05565	0.20615
0.2495	0.6618	0.12475	0.3309
0.9999	**1.6617**	0.49995	0.83085
−0.4088	**1.2529**	−0.2044	0.62645

```
float f[16]={0,
       0.8311,
      −0.2977,
       0.4961,
       0.6488,
      −0.3401,
      −0.0341,
      −0.2336,
      −0.3801,
      −0.3984,
      −0.2568,
       0.4884,
       0.1113,
       0.2495,
       0.9999,
      −0.4088};
```

FIGURE L4–1 (a) *float2.h* header file.

```
#include <regs.h>
#include "float2.h"  //Floating point data

short safe_add(short A, short B,int *status);
void rescale(short g[]);

main()
{

        short g[16];
        short sum;

        int i,k, OF, OF_LOCATION, NUM_TIMES_SCALE,temp;

        for(i=0;i<16;i++)
        {
                g[i]=0x7fff*f[i];  //Convert to Q-15, good approximate
        }

        NUM_TIMES_SCALE = 0;

start:
        sum = 0;
        //Add all values to see if OVERFLOW occurs

        for(i=0;i<16;i++)
        {
                sum = safe_add(sum,g[i],&OF);
                if(OF == 1)
                {
                        rescale(g);
                        NUM_TIMES_SCALE++;
                OF_LOCATION = i;
                goto start;
        }
```

```
                }
        }
        OF_LOCATION++;
}
void rescale(short g[])
{
        int k;
        int temp;
        //Rescale Input since it Overflows
        for(k=0;k<16;k++)
        {
                temp = (0x4000 * g[k]) << 1; //Half it
                g[k] = temp >> 16; //Half it
        }
}
```

FIGURE L4–1   (b) Overflow detection program.

```
MEMORY
{
        VECS:        o = 00000000h l=00200h /* reset & interrupt vectors*/
        IPRAM:       o = 00000200h l=0FE00h /* internal program memory */
        IDRAM:       o = 80000000h l=10000h /* internal data memory */
}

SECTIONS
{
        vectors  > VECS
        .tables  > IDRAM
        .data    > IDRAM
        .stack   > IDRAM
        .bss     > IDRAM.
        sysmem   > IDRAM
        cinit    > IDRAM
        .const   > IDRAM
        .cio     > IDRAM
        .far     > IDRAM
}
```

FIGURE L4–1   (c) Command file.

CHAPTER 6

Code Optimization

Four relatively simple modifications of assembly code can be done to generate a more efficient code. These modifications make use of the available C6x resources, such as multiple buses, functional units, pipelined CPU, and memory organization. They include: (a) using parallel instructions, (b) eliminating delays or NOPs, (c) unrolling loops, and (d) using word wide data.

Wherever possible, parallel instructions should be used to make maximum use of idle functional units. It should be noted that, whenever the order in which instructions appear is important, care must be taken not to have any dependency in the operands of the instructions within a parallel instruction. It may become necessary to have cross paths when parallelizing instructions. Cross paths allow only one source part of an instruction, say, on the A side to come from the B side. A cross path is indicated by x in the functional unit assignment. The destination is determined by the index 1 or 2. For example, consider MPY .M1X A2,B3,A4 vs. MPY .M2X A2,B3,B4.

Wherever possible, branches should be placed five places ahead of where they are intended to appear. This will create a delayed branch and minimize the number of NOPs. This approach should also be applied to load and multiply instructions that involve 4 and 1 delays, respectively. If code size is not of concern, loops should be copied, replacing any iterative portion of the code. By copying or unrolling a loop, fewer clock cycles would be needed (primarily due to deleting branches). Figure 6–1 shows the optimized version of the dot product example or `dotp` function incorporating the preceding modifications.

```
Loop:
        LDH.D1    *A8++,A2      ;load input 1 into A2
    ||  LDH.D2    *B9++,B3      ;load input 2 into B3
  [B0]  SUB.L2    B0,1,B0       ;decrement counter
  [B0]  B.S1      Loop          ;branch to Loop
        NOP       2             ;5 latency slots required
        MPY.M1X   A2,B3,A4      ;A4=A2*B3, crosspath
        NOP
        ADD.L1    A4,A6,A6      ;A6 += A4
```

FIGURE 6–1 Optimized dot product example.

76   C6x-Based DSP

```
            .def     _enable_cache

_enable_cache:

            b    .s2  B3
            mvc  .s2  CSR,B0
            clr  .s2  B0,5,7,B0
            set  .s2  B0,6,6,B0
            mvc  .s2  B0,CSR
            nop
```

FIGURE 6–2   Enabling cache feature.

Considering that there exists a delay associated with getting information from off-chip memory, whenever possible, program codes should be run from the on-chip RAM. For situations when program codes would not fit into the on-chip RAM, faster execution can be achieved by placing the main routine or function on the on-chip memory. The C6x has a cache feature that can be enabled to turn the program RAM into cache memory. This is done by setting the Program Cache Control (PCC) bits of the CSR to 010. For looping operations, it is recommended that this feature be enabled, since in this way there is a good chance that the cache will contain the needed fetch packet, and the EMIF is left unused speeding up code execution. Figure 6–2 shows the code for enabling the cache feature. The instructions CLR and SET are used to clear and set bits from the second argument position to the third argument position. For more detailed operation of cache and its options, refer to the CPU Reference Guide [1].

6.1 WORD WIDE OPTIMIZATION

If data is in halfwords (16 bits), it is possible to perform two loads in one instruction, since CPU registers are 32 bits wide. In other words, as shown in Figure 6–3, one data sample can get loaded into the lower part of a register and another one into the upper part.

However, this would require that the programmer be aware of the location of data during their manipulation. For example, to do a multiplication, two multiplication in-

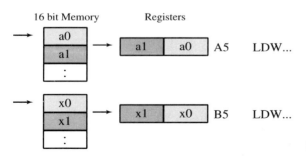

FIGURE 6–3   Use of LDW to load data.

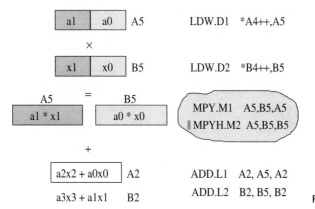

LDW.D1 *A4++,A5

LDW.D2 *B4++,B5

MPY.M1 A5,B5,A5
|| MPYH.M2 A5,B5,B5

ADD.L1 A2, A5, A2
ADD.L2 B2, B5, B2

FIGURE 6–4 Word-wide optimization.

structions, MPY and MPYH, should be used, one taking care of the lower part and the other the upper part, as is shown in Figure 6–4.

Figure 6–5 provides the word-wide optimized version of the dot product function dotp. When the looping is finished, register A2 would contain the sum of even terms and register B2 the sum of odd terms. To obtain the total sum, these registers are then added outside the loop.

```
            .def    DotP

;A4 = &a, B4 = &x, A6 = 20, B3 = return address

DotP:       zero    A2                      ;A2=0
         || zero    B2                      ;B2=0
            mv      A6,B0                   ;set B0 to argument passed in A6

loop:
            ldw     .d1     *A4++,A5        ;input word
         || ldw     .d2     *B4++,B5        ;input word
    [B0] sub        .l2     B0,1,B0         ;decrement loop counter
    [B0] b          .s1     loop            ;branch to loop (5 delay slots filled below)
            nop             2
            mpy     .m1     A5,B5,A5        ;A5=A5(low)*B5(high)
         || mpyh    .m2     A5,B5,B5        ;B5=A5(high)*B5(high)
            nop
            add     .l1     A2,A5,A2        ;A2 += A5
         || add     .l2     B2,B5,B2        ;B2 += B5

rtn:        b       .s2     B3              ;branch back to calling address
            add     .l1     A2,B2,A4        ;A4 = A2 + B2 return value
            nop             4
```

FIGURE 6–5 Word-wide optimized version of dot product code.

```
//Prototype
short DotP(int *m, int *n, short count);

//Declarations
short a[40] = {40,39,...1};
short x[40] = {1,2,...40};
short y = 0;
main()
{
        y = DotP((int *)a, (int *)x, 20);
}

short DotP(int *m, int *n, short count)
{
        short i;
        short productl;
        short producth;
        short suml = 0;
        short sumh = 0;

        for(i=0, i<count; i++)
        {
                productl = _mpy(m[i],n[i]);
                producth = _mpyh(m[i],n[i]);
                suml += productl;
                sumh += producth;
        }
        suml += sumh;
        return(suml);
}
```

FIGURE 6–6   Word-wide optimized code in C.

Out of the foregoing modifications, it is possible to do the last one, word-wide optimization, in C. This demands using an appropriate data type in C. Figure 6–6 shows the word-wide optimized C version.

6.2 SOFTWARE PIPELINING

Software pipelining is a technique for writing highly efficient assembly loop codes on C6x processors. Using this technique, all functional units on the processor are fully utilized. However, to write hand-coded software-pipelined assembly code, a fair amount of coding effort is required, due to the complexity and number of steps involved in writing such code. In particular, for complex algorithms encountered in many communications and signal/image processing applications, hand-coded software pipelining considerably increases coding time. The software tools, the compiler optimization levels 2 and 3 (**-o2** and **-o3**), attempt to achieve software pipelining to some degree without hand coding. (Refer to Figure 3–1.)

As compared with linear assembly, which is discussed next, the increase in code efficiency is rather slight. That is why hand-coded software pipelining is only briefly discussed in Lab 5. The interested reader is referred to the TI TMS320C6x Programmer's Guide [7] for more details on how to hand write software pipelined assembly code.

6.2.1 Linear Assembly

As just mentioned, software pipelining is needed to achieve maximum performance, but it takes a fair amount of coding effort to do hand-coded software pipelining. Linear assembly is a coding scheme that allows efficient codes (as compared with C) to be written with less coding effort (as compared with hand-coded software-pipelined assembly). The assembler optimizer is the software tool that parallelizes linear assembly code across the eight functional units. It attempts to achieve a good compromise between code efficiency and coding effort.

In a linear assembly code, it is not required to specify any functional units, registers, and NOPs. Figure 6–7 shows the linear assembly code version of the dot product function. The directives .proc and .endproc define the beginning and end of the linear assembly procedure. The arguments passed into and out of the procedure are indicated by registers A4, B4, A6, and B3. Register B3, referred to as a preserved register, is passed in and out with no modification. This is done to prevent it from being used by the procedure. Here this register is used to contain the return address reached by the branch instruction outside of the procedure. To preserve registers not to be used in a linear assembly procedure, such registers must be specified as input and output arguments while not being used within the procedure. The symbolic names p_m, p_n, m, n, count, prod, and sum are defined by the .reg directives. The names p_m, p_n, and count are

```
           .title   "dotp.sa"
           .def     dotp
           .sect    "code"
dotp:      .proc    A4, B4, A6, B3
           .reg     p_m, m, p_n, n, prod, sum, count
           mv       A4, p_m                ;p_m now has the address of m
           mv       B4, p_n                ;p_n now has the address of n
           mv       A6, count              ;count = the number of iterations
           mvk      0, sum                 ;sum=0

loop:      .trip    40                     ;minimum 40 iterations through loop
           ldh      *p_m++, m              ;load element of m, postincrement pointer
           ldh      *p_n++, n              ;load element of n, postincrement pointer
           mpy      m, n, prod             ;prod=m*n
           add      prod, sum, sum         ;sum += prod
[count]    sub      count, 1, count        ;decrement counter
[count]    b        loop                   ;branch back to loop

           mv       sum, A4                ;store result in return register A4
           .endproc A4, B3
```

FIGURE 6–7 Linear assembly code for dot product example.

associated with registers A4, B4, and A6 by using the move MV assignment statement. As shown here, if the number of iterations is known, a .trip directive should be used to indicate this number.

To further optimize a linear assembly code, partitioning information can be added. Such information consists of the assignment of data paths to instructions. As a result, the partitioning of loops is improved, and certain bottlenecks are avoided in the hardware resources. All the optimization methods just discussed are summarized in Appendix D as a quick reference guide.

LAB 5: REAL-TIME FILTERING

The purpose of this lab is to design and implement a finite impulse response filter on the C62 and C67. The design of the filter is done by using MATLAB. Once the design is completed, the filtering code is inserted into the sampling EVM shell program as an ISR to process live signals in real-time.

L5.1 Designing a FIR Lowpass Filter

MATLAB or filter design packages can be used to obtain the coefficients for a desired FIR filter. To make the simulation more realistic, a composite signal may be created and filtered in MATLAB. A composite signal, as shown in Figure L5–1, consisting of three sinusoids can be created by the following code:

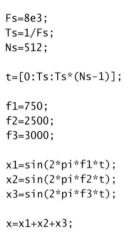

```
Fs=8e3;
Ts=1/Fs;
Ns=512;

t=[0:Ts:Ts*(Ns-1)];

f1=750;
f2=2500;
f3=3000;

x1=sin(2*pi*f1*t);
x2=sin(2*pi*f2*t);
x3=sin(2*pi*f3*t);

x=x1+x2+x3;
```

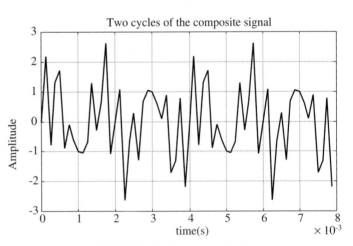

FIGURE L5–1   Composite Signal.

The signal can be better inspected by examining its frequency content. The MATLAB fft function can be used to plot the frequency content. Once done, three spikes should be observed at 750 Hz, 2500 Hz, and 3000 Hz. The frequency leakage observed on the plot is due to windowing caused by the finite observation period. A lowpass filter is designed here to filter out frequencies greater than 750 Hz and keep the lower components. The sampling frequency is chosen to be 8 KHz, which is common in voice processing. The following code is used to plot the frequency response shown in Figure L5–2:

```
X=(abs(fft(x,Ns)));
y=X(1:length(X)/2);
f=[1:1:length(y)];
plot(f*Fs/Ns,y);
grid on;
```

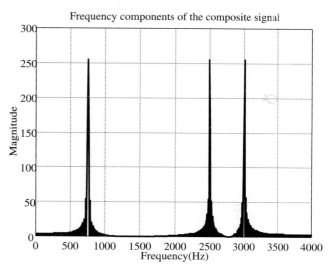

FIGURE L5–2 Frequency response of the composite signal.

To design a FIR filter with specifications of passband frequency = 1600 Hz, stopband frequency = 2400 Hz, passband gain = 0 dB, stopband attenuation = 20 dB, sampling rate = 8000 Hz, the Parks–McClellan method is used via the remez function of MATLAB [15]. The coefficients and the required MATLAB code are shown in the following code, Figure L5–3 and Table L5–1:

```
N=10;
F=[0 0.4 0.6 1];
M=[1 1 0 0];
B=remez(N,F,M);
A=1;
freqz(B,A);
```

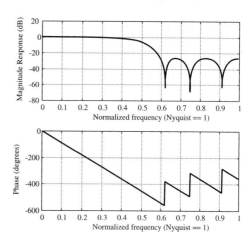

FIGURE L5–3 Filter magnitude and phase response.

Since the C62 is a fixed-point processor, numbers are represented using Q-15 format. Multiplying the coefficients or numbers with 0x7FFF results in their Q-15 representation.

82   C6x-Based DSP

TABLE L5–1   FIR Filter Coefficients

Coefficient	Values	Q-15 Representation
B0	0.0537	0x06DF
B1	0.0000	0x0000
B2	−0.0916	0xF447
B3	−0.0001	0xFFFD
B4	0.3131	0x2813
B5	0.4999	0x3FFC
B6	0.3131	0x2813
B7	−0.0001	0xFFFD
B8	−0.0916	0xF447
B9	0.0000	0x0000
B10	0.0537	0x06DF

(Note: Do not confuse B coefficients with B registers!)

Using the preceding coefficients, the filter function of MATLAB is used to verify that the FIR filter is actually able to filter out the 2.5- and 3-KHz signals. The following code allows one to visually inspect the filtering operation:

```
subplot(3,1,1);
va_fft(x,1024,8000);
subplot(3,1,2);
[h,w]=freqz(B,A,512);
plot(w/(2*pi),
   10*log(abs(h)));
grid on;
subplot(3,1,3);
y = filter(B,A,x);
va_fft(y,1024,8000);

function va_fft(x,N,Fs)
X=fft(x,N);
XX=(abs(X));
XXX=XX(1:length(XX)/2);
y=XXX;
f=[1:1:length(y)];
plot(f*Fs/N,y);
grid on;
```

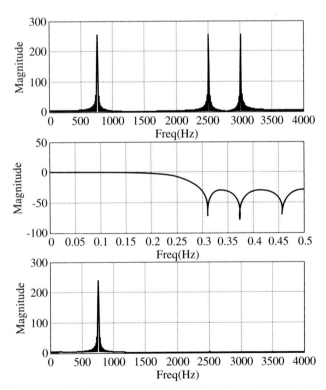

FIGURE L5–4   Frequency representation of the filtering operation.

```
n=128
subplot(2,1,1);
plot(t(1:n),x(1:n));
grid on;
xlabel('time(s)');
ylabel('Amplitude');
title('Original and
   Filtered Signals');
subplot(2,1,2);
plot(t(1:n),y(1:n));
grid on;
xlabel('time(s)');
ylabel('Amplitude');
```

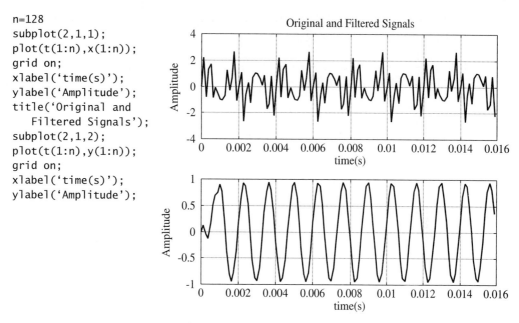

FIGURE L5-5 Time domain representation of the filtering operation.

Looking at the plots shown in Figures L5-4 and L5-5, it is clear that the filter is able to remove the right frequency components of the composite signal. Observe that the time response has an initial setup time causing the first few data samples to be inaccurate. Now that the filter design is complete, the implementation of the filter is discussed.

L5.2 Filter Implementation on the C62

A designed FIR filter can be implemented on the C62 in C or assembly. The goal of the implementation is to have an algorithm with a minimum cycle time, [i.e., to do the filtering as fast as possible to achieve the highest sampling frequency (smallest sampling time interval)]. Initially the filter is implemented in C, since this demands the least coding effort. Once a working algorithm in C is obtained, the compiler optimization levels (i.e., **-o2** and **-o3**) are activated to try and reduce the number of cycles. An implementation of the filter is then done in hand-coded assembly, which can be software pipelined for optimum performance. A final implementation of the filter is performed in linear assembly, and the timing results are compared.

The difference equation $y[n] = \sum_{k=0}^{N-1} B_k * x[n - k]$ is implemented to realize the foregoing designed FIR filter. Since the filter is implemented on the EVM, the coding is done by modifying the sampling program in Lab 3. The sampling program in Lab 3 uses an ISR that is able to receive a single word of data from the serial port and send it back out without any modification. By examining the Disassembly window as part of the debugger utility, it is possible to see that the macros MCBSP_READ/WRITE allow a word of data to be read from the DRR (0x018C0000) and then to be written into the DXR (0x018C0004). The point of concern here is that the data are 32-bits wide.

```
;----------------------------------------
temp = MCBSP_READ(0);
;----------------------------------------
    ZERO    .L1     A0
    MVKH    .S1     0x18c0000,A0
    LDW     .D1T1   *A0,A0
    NOP     4
    STW     .D2T1   A0,*+SP(4)
;----------------------------------------
MCBSP_WRITE(0, temp);
;----------------------------------------
    LDW     .D2T2   *+SP(4),B5
    NOP     2
    MVK     .S2     0x18c0004,B4
    MVKH    .S2     0x18c0004,B4
    STW     .D2T2   B5,*B4
```

Since Q-15 representation is used here, the MPY instruction cannot be used to multiply a 32-bit number with a 16-bit coefficient, as this will not give a correct result. Instead the upper part of the sampled data must be multiplied with the coefficients, which are only 16 bits wide. The result has to be left shifted by one to get rid of the extended sign bit. The number may be stored in 32 bits if a left shift of 1 is used, or in 16 bits if a right shift of 15 is used.

To perform the preceding steps in C, the _mpyhl() intrinsic and the binary shift operator << are used as shown next. Alternatively, the sample in the DRR can be right shifted by 16 and stored as a short.

```
result = ( _mpyhl(sample,coefficient) ) << 1;
```

For the proper operation of the FIR filter, it is required that the current sample and $N - 1$ previous samples be processed at the same time, where N is the number of coefficients. Hence, the N most current samples have to be stored and updated with each incoming sample. This can be done easily via the following code:

```
interrupt void serialPortRcvISR (void)
{
        int i,temp;
        temp = MCBSP_READ(0);

        //update array samples
        for(i=N-1;i>=0;i--)
                samples[i+1]=samples[i];
        samples[0] = temp;
        MCBSP_WRITE(0, temp);
}
```

This approach adds some overhead to the ISR, but, for now, it is acceptable, since at a sampling frequency of 8 KHz, there is a total of 16,625 available cycles $(1/(8\text{ KHz}/133\text{ MHz}) = 16{,}625)$ between consecutive samples, considering that the EVM here is configured to run at 133 MHz. The total overhead for this manipulation is 799 cycles without any optimization. For practical purposes, this is an acceptable number of cycles. However, it should be noted that the proper way of doing this type of filtering is by using circular buffering. This approach will be discussed in Lab 7, where hand-coded assembly is used to implement an adaptive filter.

Now that the N previous samples are in a global array, the filtering operation may be started. All that needs to be done, according to the difference equation, is to multiply each sample with the corresponding coefficient and sum the result. This is achieved by the following code:

```
interrupt void serialPortRcvISR (void)
{
        int i,temp,result = 0;
        temp = MCBSP_READ(0);

        //update array samples
        for(i=N-1;i>=0;i--)
                samples[i+1]=samples[i];
        samples[0] = temp;

        //Filtering
        for(i=0;i<=N;i++)
                result += ( _mpyhl(samples[i],b[i]) ) << 1;

        MCBSP_WRITE(0, result);
}
```

Using a function generator and an oscilloscope, it is possible to verify that the filter is working as expected. As the frequency is increased, the signal attenuation starts at 1.6 KHz and it dies out at 2.4 KHz, which agrees with the design requirements.

Considering that a working design is at hand, it is time to start the optimization of the filtering algorithm. The first step in optimization is to use the compiler optimizer. By adding **-ox** options to the command line, the optimizer is invoked without having to write or change any code. Table L5–2 summarizes the timing results for different optimization levels.

TABLE L5–2 Timing cycles for different builds.

Build Type	Number of Cycles
Compile without opimization	1581
Compile with -o0	1368
Compile with -o1	1147
Compile with -o2/-o3	421

For benchmarking purposes, all builds are done without the **-g** option. By typing `go serialPortRcvISR`, it is possible to locate the beginning of the ISR. Adding a breakpoint there provides the beginning point of the benchmark. The ISR branches back to the while loop with the statement B IRP (seen in the Disassembly window). By adding a second breakpoint there, the endpoint of the benchmark is defined. Entering the `runb` command starts the benchmarking and activates the `clk` variable, allowing the user to determine the number of cycles it takes for the ISR to complete. It is usually difficult to be totally accurate with the timing results, so an extra 5 cycles may be added to the result to accommodate for the last 5 cycles as the ISR is branching back. Notice that these timings include the receive and send of the sample, the array sorting and the filtering operation.

Before doing linear assembly, the algorithm is first written in assembly to see how basic optimization methods such as placing instructions in parallel, filling delay slots, loop unrolling and word-wide optimization affect the timing cycle of the algorithm.

To perform the operation of multiplying and adding N coefficients, a loop has to be set up. This can be done by using a branch instruction. A counter is required to exit the loop once N iterations have been performed. For this purpose, one of the conditional registers (A1, A2, B0, B1, or B2) is used. Any other register does not allow for conditional testing. Adding [A2] in front of an instruction allows the processor to execute the instruction if the value in A2 does not equal zero. If A2 equals zero, the instruction is skipped. (Note that an instruction cycle is still consumed.) The S1 unit may be used to perform the move constant and branch operations. The value in the conditional register A2 is decremented by using a subtract instruction. Since the subtract operation should stop if the value drops below zero, this conditional register is included in the SUB instruction to execute it only if the value is not equal to zero. It should be remembered to add 5 delay slots for the branch instruction. This loop is now shown:

```
            MVK     .S1     11, A2      ;move 11 into A2 count register
Loop1:
            .
            .
            .

    [A2]    SUB     .L1     A2,1,A2     ;decrement counter
    [A2]    B       .S1     Loop1       ;branch back to Loop1
            NOP     5
```

We can now start adding instructions to perform the multiplication and accumulation of the values. First, those values that are to be multiplied need to be loaded from their memory locations into the registers. This is done by using load word (LDW) and load half word (LDH) instructions. Upon executing the load instructions, the pointer is postincremented so that it is pointed to the next memory value. Once the values have appeared in the registers (four cycles after the load instruction), the MPYHL instruction is used to multiply the two values and store the result in another register. Then, the summation is performed by using the ADD instruction. The following is the completed assembly program:

```
            .global    _fir_simple
_fir_simple:
            MV         .S1        A6,A2           ;Count register
            ZERO       .S1        A9              ;Sum register

loop:       LDW        .D1        *A4++,A7        ;Load data from samples
            LDH        .D2        *B4++,B7        ;Load data from coefficients
            NOP        4
            MPYHL      .M1x       A7,B7,A8        ;A7 is 32 bit sample, B7 is Q-15
                                                   representation coefficient
            NOP
            SHL        A8,1,A8                    ;Eliminate sign extension bit
            ADD        .S1        A8,A9,A9        ;Accumulate result
    [A2]    SUB        .S1        A2,1,A2         ;Decrement counter
    [A2]    B          .S1        loop
            NOP        5

            MV         .S1        A9,A4           ;Move result to return register
            B          .S2        B3              ;Branch back to calling address
            NOP        5
```

The supplied code is a C-callable assembly function. To call this function from C, a function declaration must be added as external (extern) without any arguments. The parameters to the function are passed via registers A4, B4, and A6. The return value is stored in A4. Here the pointers to the arrays are passed in A4 and B4 as the first two arguments and the number of iterations in A6 as the third argument. The return address from the function is stored in B3. Therefore, a final branch to B3 is required to return from the function. For a complete explanation of calling assembly functions from C, see the TI TMS320C6x Optimizing C Compiler User's Guide [5].

Running this code results in a total of 1423 cycles for the ISR. 570 out of 1423 cycles is the number of cycles it takes for the assembly function to complete. Note that here the assembly function should complete in only 192 cycles. The reason for the difference is that the EVM is running the code from the external memory in the SBSRAM. To move the code so that it runs in the internal program memory, a section in the command file should be defined. The assembly file must also contain the directive .sect with the same name so that the linker would know which part of the code to place in the internal memory. By adding the directive .sect "fir" at the beginning of the assembly file, it is possible to move the code into the internal memory space. Running the code from there results in 192 cycles, as expected. Notice that only the assembly function is running in the internal program memory and that the rest of the ISR is still slow taking a total of 1020 cycles.

It is possible to move the complete ISR into the internal memory space to allow a faster execution of the interrupt. This can be performed by using the principle just discussed.

To optimize the preceding function, basic optimization methods, such as placing instructions in parallel, filling delay slots, and loop unrolling, are used. By examining the code, it can be seen that some of the instructions can be placed in parallel. Care must be taken not to schedule instructions in parallel that use previous operands as their operands because of operand dependencies. The two initial load instructions are independent and can be made to run in parallel. By looking at the rest of the program, it can be seen that the rest of the operands are dependent on the previous operands, and hence no other instructions can be placed in parallel.

To reduce the cycles taken by the NOP operations, it is possible to use a technique known as filling delay slots. For example, as the load instructions are executed in parallel, it is possible to schedule the subtraction of the loop counter in place of their NOPs. The branch instruction takes five cycles to execute. It is also possible to slide the branch instruction four slots up to get rid of its NOPs. Incorporating these optimizations, the function is rewritten as follows:

```
             .global   _fir_filled

             .sect     ".fir_filled"          ;used to load into internal
                                               program memory
_fir_filled:
             MV        .S1       A6,A2         ;Count register
             ZERO      .S1       A9            ;Sum register

loop:        LDW       .D1       *A4++,A7      ;Load data from samples
          || LDH       .D2       *B4++,B7      ;Load data from coefficients
             NOP
[A2]         SUB       .S1       A2,1,A2       ;Decrement counter
[A2]         B         .S1       loop          ;branch back to loop
             NOP
             MPYHL     .M1x      A7,B7,A8      ;A7 is 32 bit sample, B7 is Q-15
                                                representation coefficient
             NOP
             SHL                 A8,1,A8       ;Eliminate sign extension bit
             ADD       .S1       A8,A9,A9      ;Accumulate result


             MV        .S1       A9,A4         ;Move result to return register
             B         .S2       B3            ;Branch back to calling address
             NOP       5
```

By filling delay slots, the number of cycles is reduced within the loop to $9 \times 11 = 99$. In repetitive loops such as this one, it can easily be seen that the branch instruction takes up extra cycles that can be eliminated. As just mentioned, one method to do this elimination is to fill delay slots by sliding the branch instruction higher in the execution phase, thus filling the latencies associated with this instruction. Another method to reduce the effect of the latencies is to unroll the loop. However, notice that loop unrolling only eliminates the last latency of the branch. Since in the delay-filled version, the branch latency has no effect on the number of cycles, loop unrolling does not achieve any further timing improvement.

To perform word optimization, the ISR has to be modified to store a sample into 15 bits rather than 32. This is achieved by simply shifting the input right by 15 bits during the read process. The followiing code stores a sample into a short variable, assuming that the output result is 32 bits:

```
interrupt void serialPortRcvISR (void)
{
        int i,result = 0;
        short temp;
        temp = MCBSP_READ(0) >> 15;
        //Filtering
        MCBSP_WRITE(0, result);
}
```

Using word optimization, it is possible to load two consecutive 15-bit values in memory with a single load-word instruction. The register that the word gets loaded to will contain the two values in the lower and upper parts. The instructions MPY and MPYH can then be used to multiply the upper and lower parts, respectively. The following assembly code shows how this is done for the FIR filtering program:

```
            .global   _fir_wordoptimized
_fir_wordoptimized:
            MV        .S1    A6,A2          ;Count register
            ZERO      .S1    A9             ;Sum register
    ||      ZERO      .S2    B9

loop:       LDW       .D1    *A4++,A7       ;Load data from samples (here
                                             the input data is in 15bit
                                             format)
    ||      LDW       .D2    *B4++,B7       ;Load data from coefficients
   [A2]     SUB       .S1    A2,1,A2        ;Decrement counter
   [A2]     B         .S1    loop
            NOP       2
            MPY       .M2    A7,B7,B8       ;B8 is the lower part product
    ||      MPYH      .M1    A7,B7,A8       ;A7 is the higher part product
            NOP
            ADD       .S1    A8,A9,A9       ;Accumulate result
    ||      ADD       .S2    B8,B9,B9       ;Accumulate result

            LDH       .D1    *A4++,A7       ;Load the final elements
    ||      LDH       .D2    *B4++,B7       ;Load the final elements
            NOP       4
            MPY       .M1    A7,B7,A8       ;Final multiply
            NOP
            ADD       .L1    A8,A9,A9       ;Final add

            ADD       .S1    A9,B9,A4       ;Move result to return register
            SHL              A4,1,A4        ;Eliminate sign extension bit

            B         .S2    B3             ;Branch back to calling address
            NOP       5
```

Notice here that since two loads are done consecutively, it takes half the amount of time to loop through the program. When calling this function, the value passed in A6 must be $N/2$, where N is the number of coefficients of the FIR filter. In our case, we have 11 coefficients and need to perform an additional multiply and accumulate, as shown in the foregoing code. With this code, it is possible to bring down the number of cycles to 51. The timing cycles for the above mentioned optimizations are shown in the following table:

Optimization	Cycles
Un-optimized assembly	192
Delay slot filled assembly	104
Word optimized assembly	51

L5.2.1 Hand Written Software-Pipelined Assembly

To produce a software-pipelined version of the algorithm, it is required to first write it in symbolic form without any latency or register information. The following code shows how to write the FIR algorithm in a symbolic form:

```
          LDW     *p_sample++,sample    ;load sample word
          LDH     *p_coef++,coef        ;load coef half-word
          MPYHL   sample,coef,temp      ;temp = sample(high)*coef
          SHL     temp,1,temp           ;shift left to remove sign extended bit
          ADD     sum,temp,sum          ;sum += temp
[count]   SUB     count,1               ;decrement counter
[count]   B       loop                  ;branch back to loop
```

To hand-write software-pipelined code, a dependency graph of the algorithm must be drawn, and a scheduling table must be created from it. The code for the algorithm is then derived from the scheduling table. In a dependency graph (see Figure L5–6 for the termi-

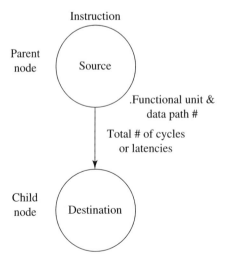

FIGURE L5–6 Dependency graph terminology.

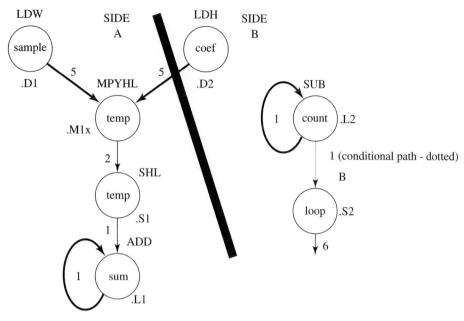

FIGURE L5–7   FIR dependency graph.

nology), the nodes denote instructions and symbolic variable names. The paths show the flow of data and are annotated with the latencies of their parent nodes. To draw a dependency graph, we start by drawing nodes for the instructions and symbolic variable names.

After the basic dependency graph is drawn, functional units have to be assigned. A line is drawn between the two sides of the CPU so that the workload is split as equally as possible. It is apparent that the loads should be done one on each side, so this provides a good starting point. From there, it is up to the programmer to choose which side to place the rest of the instructions to equally divide the workload between the A side and B side functional units. The completed dependency graph for the FIR algorithm is shown in Figure L5–7.

The next step for hand-writing a pipelined code is to setup a scheduling table. To do so, the longest path must be identified in order to determine how long the table should be. Counting the latencies of each side, it is seen that the longest path on the left side is 8 and on the right side 7. Thus, eight prolog columns are required in the table before entering the main loop. There need to be eight rows (one for each functional unit) and nine columns in the table. The scheduling is started by placing the parallel load instructions at slot zero. The instructions are repeated at every iteration thereafter. The multiply instruction must appear four slots after the loads, so it is scheduled into slot five. The shift must appear two slots after the multiply, and the add must appear after the shift instruction, making it placed in slot eight, which is the loop part of the code. The branch instruction is scheduled in slot three by reverse counting five cycles back from the loop. The subtraction must occur before the branch, so it is scheduled in slot two. The completed scheduling table appears in Table L5–3.

TABLE L5–3 FIR scheduling table

	PROLOG								LOOP
	0	1	2	3	4	5	6	7	8
.L1									ADD
.L2			SUB	SUB	SUB	SUB	SUB	SUB	SUB
.S1								SHL	SHL
.S2				B	B	B	B	B	B
.M1						MPYHL	MPYHL	MPYHL	MPYHL
.M2									
.D1	LDW	LDW	LDW	LDW	LDW	LDW	LDW	LDW	LDW
.D2	LDH	LDH	LDH	LDH	LDH	LDH	LDH	LDH	LDH

The code is hand-written directly from the scheduling table, as eight parallel instructions before entering a loop that completes all the adds. With the following code, the number of loop cycles is reduced to 20:

```
            .global     _fir_pipelined

_fir_pipelined:
            ZERO        A8
            ZERO        A9
            MV          A6,B2

            LDW         .D1     *A4++,A7
||          LDH         .D2     *B4++,B7

            LDW         .D1     *A4++,A7
||          LDH         .D2     *B4++,B7

            LDW         .D1     *A4++,A7
||          LDH         .D2     *B4++,B7
||  [B2]    SUB         .L2     B2,1,B2

            LDW         .D1     *A4++,A7
||          LDH         .D2     *B4++,B7
||  [B2]    SUB         .L2     B2,1,B2
||  [B2]    B           .S2     loop10

            LDW         .D1     *A4++,A7
||          LDH         .D2     *B4++,B7
||  [B2]    SUB         .L2     B2,1,B2
||  [B2]    B           .S2     loop10

            LDW         .D1     *A4++,A7
||          LDH         .D2     *B4++,B7
||  [B2]    SUB         .L2     B2,1,B2
||  [B2]    B           .S2     loop10
||          MPYHL       .M1     A7,B7,A8
```

```
                    LDW     .D1     *A4++,A7
         ||         LDH     .D2     *B4++,B7
         || [B2]    SUB     .L2     B2,1,B2
         || [B2]    B       .S2     loop10
         ||         MPYHL   .M1     A7,B7,A8

                    LDW     .D1     *A4++,A7
         ||         LDH     .D2     *B4++,B7
         || [B2]    SUB     .L2     B2,1,B2
         || [B2]    B       .S2     loop10
         ||         MPYHL   .M1     A7,B7,A8
         ||         SHL     .S1     A8,1,A8

loop10:             LDH     .D1     *A4++,A7
         ||         LDH     .D2     *B4++,B7
         || [B2]    SUB     .L2     B2,1,B2
         || [B2]    B       .S2     loop10
         ||         MPYHL   .M1     A7,B7,A8
         ||         SHL     .S1     A8,1,A8
         ||         ADD     .L1     A8,A9,A9

                    MV      .L1     A9,A4
                    B       .S2     B3
                    NOP     5
```

L5.2.2 Assembler Optimizer Software-Pipelined Assembly

Since hand-written pipelined codes are time consuming to write, linear assembly is usually used to generate pipelined codes. In linear assembly, latencies, functional units, and register allocations do not need to be specified. Instead, symbolic variable names are used to write a sequential code without delay slots (NOPs). The file extension for a linear assembly file is .sa. The c16x command is used to invoke the assembly optimizer. It is still possible to write a C-callable function and pass variables to it by using the same principles stated for an assembly function. However, the assembly optimizer must be allowed to create the object file before it is used.

When writing code in linear assembly, the assembler optimizer must be supplied with the right kind of information to perform optimizations. The first of these is to specify where it should optimize. The optimizer only considers code between statements .proc and .endproc. Symbolic variable names are used to allow the optimizer to select which registers to use. This is done by using the .reg directive together with the names of the variables. Also, registers that contain input arguments, such as variables passed to a function, must be specified. The registers declared to contain input arguments cannot be modified and have to be declared as arguments of the .proc statement. Registers that contain output values upon exit of the procedure must be declared as arguments of the .endproc directive.

To write the FIR algorithm in linear assembly, we start by creating the main loop and adding the load, multiply, and add instructions. Since two pointers to two arrays and an integer are passed to the function, it is required to declare registers A4, B4, and A6 as part of the `.proc` directive. Also, register A4 is used for returning values, so it appears as part of the `.endproc` directive. The preserved register B3 is indicated as an argument in both of these directives. To connect the symbolic variable names to the input registers, the move statement is used. Finally, the optimizer is told that the loop is to be performed a minimum of 11 times by inserting the `.trip` directive. Using the following code, a timing outcome of 29 cycles is obtained:

```
             .global _fir_la
             .sect ".fir_la"

_fir_la:    .proc       A4,B4,A6,B3
            .reg        p_m,m,p_n,n,prod,sum,cnt

            mv          A4, p_m         ;move argument in A4 to p_m
            mv          B4, p_n         ;move argument in B4 to p_n
            mv          A6, cnt         ;set up counter (third argument)
            zero        sum             ;sum=0

loop8:      .trip       11              ;minimum 11 times through loop
            ldw         *p_m++, m       ;load m
            ldh         *p_n++, n       ;load n
            mpy         m,n,prod        ;prod = m * n
            shl         prod,1,prod     ;prod << 1
            add         prod,sum,sum    ;sum += prod

   [cnt]    sub         cnt,1,cnt       ;decrement counter
   [cnt]    b           loop8           ;branch back to loop8

            mv          sum,A4          ;move result into return register A4
            .endproc    A4,B3

            B           B3              ;branch back to address stored in B3
            NOP         5
```

To summarize the programming approach (see Appendix D: Quick Reference Guide), it is appropriate to start writing your code in C and then use the optimizer to achieve a faster code. If the code is not as fast as expected, you may write it in assembly and incorporate the discussed simple optimization techniques. However, it is usually easier and more efficient to rewrite your code in linear assembly, since the assembly optimizer attempts to create pipelined code for you. If at the end none of these approaches provide a satisfactory timing cycle, you are left no choice but to rewrite your code in hand-coded pipelined assembly.

L5.3 Filter Implementation on the C67

Implementing the FIR filter on the C67 takes relatively less effort. Since the hardware is capable of multiplying and adding floating point numbers, Q-format is not required.

However, it is slower, since floating point operations have more latencies than their fixed point counterparts. As shown next, the FIR filter interrupt and subroutine are modified to run on a C67 EVM. The code written in C is fairly simple. The coefficients are entered directly as type float. The data buffer is declared as float, and a sample is initially read as an int then typecast to a float.

Using the -k option to keep the assembly file, it can be verified that the compiler is actually using the MPYSP and ADDSP instructions to perform the floating-point multiply and add rather than calling a separate function to do them in software.

```
float dotp1(const float a[], const float b[])
{
        int i;
        float sum = 0.0;
        for(i=0; i<11;i++)
                sum += a[i] * b[i];
        return sum;
}
interrupt void serialPortRcvISR (void)
{
        int i,sample;
        float temp, sum;
        sample = MCBSP_READ(0);
        temp = (float)sample;
        for(i=10;i>=0;i--)
                x[i]=x[i-1];
        x[0]=temp;
        sum = dotp1(coef, x);
        MCBSP_WRITE(0, sum);
}
```

A complete listing of the programs for this lab appears in Appendix A.

CHAPTER 7

Frame Processing

7.1 TRIPLE BUFFERING

When it comes to processing frames of data (for example, in doing FFT and block convolution), triple buffering is an efficient data-handling mechanism that makes use of the DMA. While samples of the current frame are being collected by the CPU in an INPUT array via an ISR, samples of the previous frame in an INTERMEDIATE array can get processed during the time left between samples. At the same time, the DMA can be used to send out samples of the two previous frame already processed and available in an OUTPUT array. In this manner, the CPU is used to set up the INPUT array and process the INTERMEDIATE array while the DMA is used to move processed data from the OUTPUT array. At the end of each frame or the start of a new frame, the roles of these arrays are interchanged. The INPUT array is reassigned as the INTERMEDIATE array to be processed, the processed INTERMEDIATE array is reassigned as the OUTPUT array to be sent out, and the OUTPUT array is reassigned as the INPUT array to collect incoming samples for the current frame. This process is illustrated in Figure 7–1.

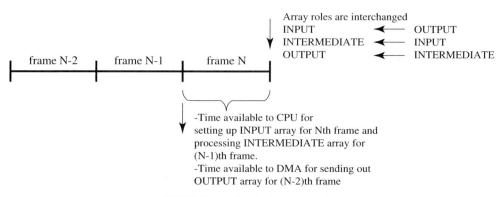

FIGURE 7–1 Triple-buffering technique.

7.2 DIRECT MEMORY ACCESS

Many DSP chips are equipped with a Direct Memory Access (DMA) resource acting as a co-processor to move data from one part of memory into another without interfering with the CPU operation. As a result, the chip throughput is increased, since the CPU and DMA can process and move data without interfering with each other. The C6x DSP has six DMA channels. Each channel has its own memory-mapped control registers that can be set up to move data from one place in memory to another. These registers contain the information regarding source and destination locations in memory, number of transfers, and format of transfers.

It is possible to transfer a block of data consisting of a number of frames that, in turn, consist of a number of elements. Elements here mean the smallest piece of data. The example shown in Figure 7–2 illustrates the DMA register setup for transferring a block of data (you can view this as image data) consisting of four frames (rows), while each frame consists of four elements (16-bit pixels). For details on setting up DMA control registers, refer to the TI TMS320C6x Peripherals Reference Guide [3].

LAB 6: FAST FOURIER TRANSFORM

Operations such as the discrete Fourier transform (DFT) or FFT require that a frame or block of data be present at the time of processing. Unlike filtering, where operations are done on every incoming sample, in frame processing, N samples are captured first, and then operations are performed on all N samples.

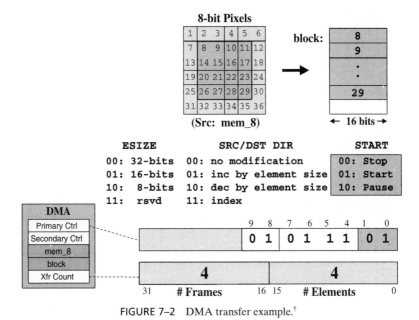

FIGURE 7–2   DMA transfer example.[†]

To perform frame processing, a proper method of gathering, processing, and sending out data is required. The processing of a frame of data is not usually completed within the sampling time interval, rather it is spread over the duration of a frame before the next frame of data is gathered. Hence, incoming samples must be stored to a separate buffer than the one being processed. Also, a separate buffer is needed to send out a previously processed frame of data. This brings forth the idea of triple buffering involving three buffers: INPUT, INTERMEDIATE, and OUTPUT. The input buffer is used to gather data using the interrupt driven subroutine. As the INPUT buffer is being filled, the data in the INTERMEDIATE buffer is being processed, and the data in the OUTPUT buffer is being sent out via the DMA, so that the CPU is not slowed to send the data out.

To do triple buffering on the C6x, the shell sampling program in Lab 3 is modified to incorporate an endless loop revolving around the rotation of three buffers. The buffers rotate every time the INPUT buffer is full, so that the new frame of N sampled data is passed to the INTERMEDIATE buffer for processing, and the previous processed frame is passed to the OUTPUT buffer for transmission. The following modifications of the shell program achieves this:

```
short *output;            /* POINTER TO DATA ARRAY FOR OUTPUT   */
short *input;             /* POINTER TO DATA ARRAY FOR INPUT    */
short *intermediate;      /* POINTER TO DATA ARRAY FOR DMA ACCESS */
int index=0;

main()
{
        int i,k;
        evm_init();
        mcbsp_drv_init();
        dev = mcbsp_open(0);
        init_serial();
        init_arrays();

        /* Main Loop, wait for Interrupt */
                for( ; ; )
        {
                wait_buffer();     /* WAIT FOR A NEW BUFFER OF DATA*/
        }
}
```

The preceding code shows how an endless loop is added to the shell program. Here, most of the initializations for the codec and McBSP have not been shown to make the code easier to follow. Once the serial port is initialized, the three arrays are allocated in memory and initialized to zero. The program then goes into an endless loop where the function `wait_buffer` is executed endlessly:

```
void wait_buffer(void)
{
        short *p;
        /* WAIT FOR ARRAY INDEX TO BE RESET TO ZERO BY ISR         */
        while(index);

        /* ROTATE DATA ARRAYS                                      */
        p = input;
        input = output;
        output = intermediate;

   //Function call here...

        intermediate = p;
        WriteFIFO_DMA();
        while(!index);
}
```

The function `wait_buffer()` checks on the global variable `index` to do the rotation of the arrays and the start of processing. When the INPUT array becomes full (indicated by `index`), the arrays are rotated and the INTERMEDIATE array gets set for processing. The comment `//Function call here...` indicates where the processing function should be placed. The ISR is modified as shown in the next block of code. Note that `index` is incremented within the ISR.

```
interrupt void serialPortRcvISR (void)
{
        int temp;
        temp = MCBSP_READ(0);

        input[index] = temp >> 16;

        MCBSP_WRITE(0, output[index] << 16);

        if (++index == BUFFLENGTH)
                index = 0;
}
```

The ISR reads an input sample from the DRR and shifts it by 16 bits since the representation of numbers are considered to be in Q-15 format. The variable `index` is incremented until BUFFLENGTH is reached.

The function `WriteFIFO_DMA()` uses the C6x's DMA capability to send the OUTPUT array to the host through the FIFO (first in first out) on the EVM. The following is the code for doing so. The two DMA API functions `dma_reset()` and `dma_init()` are used to initialize the DMA for data transfer to the FIFO. A single frame of samples of length 128 is transferred, as indicated by the content of the transfer counter register (TCR).

```
void WriteFIFO_DMA(void)
{
        dma_reset();
        dma_init(2,                              //Channel
                0x0A000110u,                     //Primary Control Register
                                                 //    (Peripherals pp4-9)

                0x0000000Au,                     //Secondary Control Register
                (unsigned int) intermediate,     //Source Address
                0x01710000u,                     //Destination Address
                0x00010080u);                    //Transfer Counter Register
        DMA_START(DMA_CH2);
}
```

With these modifications, the sampling program can be observed to work when using a function generator and an oscilloscope. A simple program called *gui.exe* is written utilizing the `evm6x_read()` function from the EVM host support library to get the data from the FIFO on the host side and display it on the screen. This program placed on the attached CD was created in Microsoft Visual C++ using a Dialog Wizard. The program basically starts a thread that continuously reads the FIFO on the host side and plots the results on the screen. If desired, this program may be upgraded to provide a better GUI.

L6.1 DFT Implementation

DFT can be simply calculated from

$$X[k] = \sum_{n=0}^{N-1} x[n] * W_N^{nk}, \quad k = 0, 1, \ldots, N-1,$$

where $W_N = e^{-j2\pi/N}$.

This equation requires N complex multiplications and $N-1$ complex additions for each term. For all N terms, N^2 complex multiplications and $N^2 - N$ complex additions are needed. As is well known, this method is not efficient, since the symmetry properties of the transform are not utilized. However, it is useful to implement this equation as a comparison to the FFT implementation. This implementation is done by using the Code Composer Studio. The graphing capability of this tool is quite handy when working with offline signals. The reason for working in an offline manner is the amount of time required to do DFT (unoptimized in C), which exceeds the duration of a frame.

First, a simple composite signal is generated in MATLAB with the frequency components located at 750 Hz, 2500 Hz, and 3000 Hz. Saving two periods of this signal sampled at 8000 Hz results in a 64-point signal. Figure L6–1 shows the signal read into the Code Composer Studio and plotted using its graphing capability. This capability is also used to plot the frequency content of the signal based on a built-in FFT algorithm.

The DFT algorithm used is the one appearing in the TI Application Report SPRA291 [16]. This algorithm is now shown:

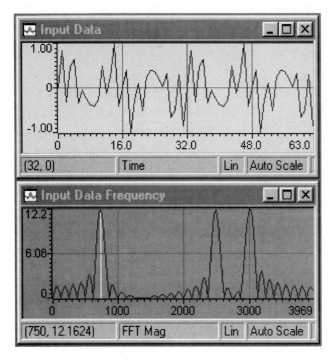

FIGURE L6–1 Input signal.

```
#include <math.h>
#include "params.h"
void dft(int N, COMPLEX *X){
        int n, k;
        double arg;
        int Xr[1024];
        int Xi[1024];
        short Wr, Wi;
        for(k=0; k<N; k++){
                Xr[k] = 0;
                Xi[k] = 0;
                For(n=0; n<N; n++){
                        arg =(2*PI*k*n)/N;
                        Wr = (short)((double)32767.0 * cos(arg));
                        Wi = (short)((double)32767.0 * sin(arg));
                        Xr[k] = Xr[k] + X[n].real * Wr + X[n].imag * Wi;
                        Xi[k] = Xi[k] + X[n].imag * Wr - X[n].real * Wi;
                }
        }

        for (k=0;k<N;k++){
                X[k].real = (short)(Xr[k]>>15);
                X[k].imag = (short)(Xi[k]>>15);
        }
}
```

102 C6x-Based DSP

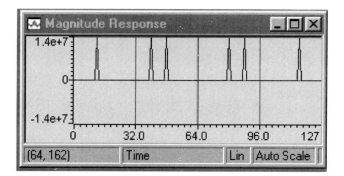

FIGURE L6–2 Magnitude response of DFT.

In order to use this algorithm, the input has to be represented as complex numbers. This is done using a `struct` definition to create a complex variable with components `real` and `imag`. Below is the main program used to perform DFT.

```
main()
{
        int i,j;
        COMPLEX x[128];
        int mag[128];

        /*Change input to Q-15*/
        for(i=0;i<128;i++)
        {
                x[i].real=0x7FFF * input_data[i];
                x[i].imag=0;
        }
        dft(128, x);

        for(i=0;i<128;i++)
                mag[i]=(x[i].real*x[i].real + x[i].imag*x[i].imag) << 1;
        return(0);
}
```

The input is converted to Q-15 format and stored in the complex structure, which is then used to call the DFT function. The magnitude of the DFT outcome is shown in Figure L6–2. As expected, there are three spikes: one at 750 Hz, one at 2500 Hz, and one at 3000 Hz. Notice that this code is quite inefficient, as it calculates each twiddle factor using the math library at every iteration. Running this code from the external SB-SRAM results in an execution time of about 1.8×10^9 cycles for a 128-point frame.

L6.2 FFT Implementation

The preceding DFT code will not run in real-time on the EVM, since there are only $16{,}625 \times N$ cycles to perform an N-point transform. For a 128-point signal, the limit is about 2×10^6 cycles, and the preceding timing exceeds this limit.

To make use of the symmetry properties of the transform, the approach of computing a 2N-point FFT, as mentioned in the TI Application Report SPRA291 [16], is used. This approach involves forming two new signals $x_1[n]$ and $x_2[n]$ from the input signal $x[n]$ by splitting it into even and odd parts as follows:

$$x_1[n] = x[2n], \quad 0 \leq n \leq N - 1,$$
$$x_2[n] = x[2n + 1].$$

From the two sequences $x_1[n]$ and $x_2[n]$, a new complex sequence is defined as

$$\hat{x}[n] = x_1[n] + jx_2[n], \quad 0 \leq n \leq N - 1$$

To get $X[k]$, the following equations are then used:

$$\begin{cases} X[k] = \hat{X}[k]A[k] + \hat{X}^*[N - k]B[k], & k = 0, 1, \ldots, N - 1, \\ X[N] = X[0], X[2N - k] = X^*[k], & k = 0, 1, \ldots, N - 1, \end{cases}$$

where

$$A[k] = \frac{1}{2}(1 - jW_{2N}^k),$$

$$B[k] = \frac{1}{2}(1 + jW_{2N}^k).$$

As a result, a 2N-point transform is calculated based on an N-point transform, which leads to a reduction in the number of cycles. The codes for the functions (`split1`, `R4DigitRevIndexTableGen`, `digit_reverse`, `radix4`) implementing this approach are provided in the TI Application Report [16] and appear in Appendix B.

Figure L6–3 shows the FFT results where the signal has been scaled down zero, two, four, and five times, respectively. The scaling is done to get rid of overflow, as seen for the scale factors zero, two, and four. As revealed by these figures, the input signal has to be scaled down five times to eliminate the effects of overflow. When the signal is scaled down five times, the expected peaks appear.

The total number of cycles for the FFT is 224,512. Since this is less than the time available for a 128-point data frame at a sampling rate of 8 KHz, it is expected that this algorithm would run in real-time on the C6x processor.

L6.3 Real-Time FFT

To perform the FFT in real-time, the triple-buffering program is used. A frame length of 128 is considered here. The output is observed by halting the processor through the Code Composer Studio. The animate feature of the Code Composer Studio cannot be used here, since it slows down the processing and causes frames to overlap.

The following modifications are made to the triple buffering program to run the FFT algorithm in real-time:

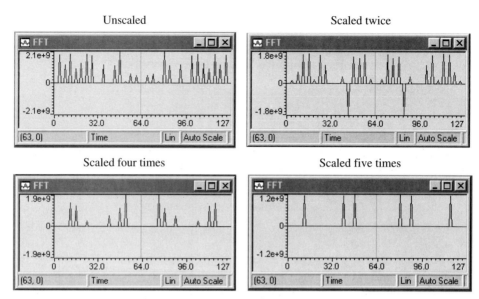

FIGURE L6–3   FFT magnitude response.

```
void wait_buffer(void)
{
        int n,k;
        short *p;

        while(index);

        p = input;
        input = output;
        output = intermediate;

        for (n=0; n<NUMPOINTS; n++)
        {
                x[n].imag = p[2*n + 1];   // x2(n) = g(2n + 1)
                x[n].real = p[2*n];       // x1(n) = g(2n)
        }

        radix4(NUMPOINTS, (short *)x, (short *)W4);
        digit_reverse((int *)x, IIndex, JIndex, count);
        x[NUMPOINTS].real = x[0].real;
        x[NUMPOINTS].imag = x[0].imag;

        split1(NUMPOINTS, x, A, B, G);
        G[NUMPOINTS].real = x[0].real - x[0].imag;
        G[NUMPOINTS].imag = 0;
        for (k=1; k<NUMPOINTS; k++){
                G[2*NUMPOINTS-k].real = G[k].real;
                G[2*NUMPOINTS-k].imag = -G[k].imag;
```

```
            }
            for (k=1; k<NUMDATA; k++){
                    mag1[k] = (G[k].real*G[k].real) << 1;
                    mag2[k] = (G[k].imag*G[k].imag) << 1;
                    mag[k] = mag1[k] + mag2[k];
            }
            intermediate = p;
            WriteFIFO_DMA();
            while(!index);
    }
```

The wait_buffer() function is modified with the required function calls so that when the input buffer is full, the FFT is calculated and sent out to the host through the FIFO. Several other initializations are done in order to get the algorithm running in real-time. These include the initialization of the coefficients and the twiddle factors.

The functionality of the code can be verified by connecting a function generator to the line-in port and running the code. The graphing capability of the Code Composer Studio can be employed to plot arrays directly. By changing the frequency of the input, the spikes in the frequency response would move to left or right accordingly. Figure L6–4 illustrates the output for a 1-KHz and 2-KHz sinusoidal signal. These snap shots are captured by halting the processor. Here the input is scaled by shifting it right 20 bits.

A complete listing of the programs for this lab appear in Appendix B.

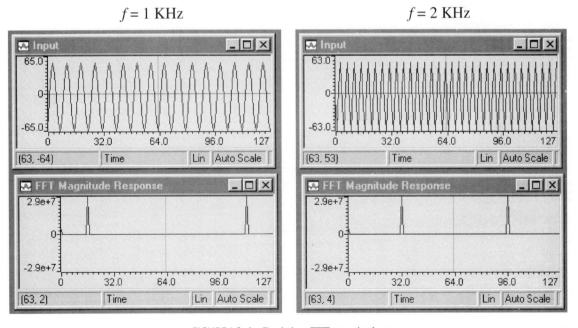

FIGURE L6–4 Real-time FFT magnitude response.

CHAPTER 8

Circular Buffering

In many DSP algorithms, such as adaptive filtering or spectral analysis, we need to deal with a moving window. Circular buffering is an addressing mode by which a moving-window effect can be easily created. In a circular buffer, if a pointer pointing to the last element of the buffer is incremented, it is wrapped around and pointed back to the first element of the buffer. This provides an easy mechanism to exclude the oldest sample while including the newest sample creating a moving-window effect as illustrated in Figure 8–1.

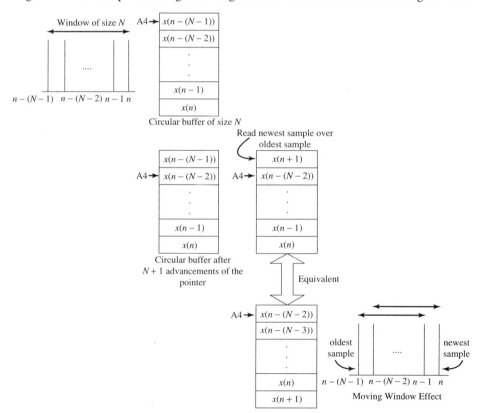

FIGURE 8–1 Moving-window effect.

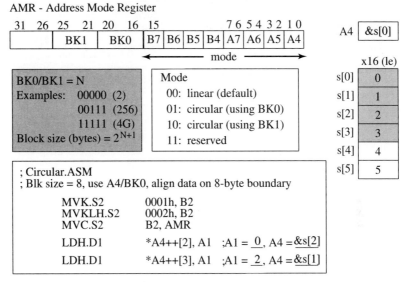

FIGURE 8–2 Setting up a circular buffer.[†]

Some DSPs have dedicated hardware for doing this type of addressing. On the C6x processor, the arithmetic logic unit has the circular buffer addressing capability built into it. To use circular buffering, first the buffer size needs to be written into appropriate bits of the Address Mode Register (AMR). Then, the register to be used as the circular buffer pointer needs to be specified by setting appropriate bits of the AMR. The C6x allows two independent circular buffers of powers of two sizes. Care must be taken to align data on the buffer size boundary. Figure 8–2 gives the code for setting up the AMR register for a circular buffer of size eight.

LAB 7: ADAPTIVE FILTERING

Adaptive filtering is used in many applications ranging from noise cancellation to system identification. In most cases, the coefficients of an FIR filter are modified according to an error signal in order to adapt to a desired signal. In this lab, a system identification example is implemented where an adaptive FIR filter is used to adapt to the output of a seventh-order IIR bandpass filter. The IIR filter is designed in MATLAB and implemented on the C62 EVM in C. The adaptive FIR is first implemented in C and later in assembly using circular buffering.

In system identification, the behavior of an unknown system is modeled by accessing its input and output. An adaptive FIR filter can be used to adapt to the output of the system based on the same input. The difference in the output of the system $d[n]$ and the output of the adaptive filter $y[n]$ constitutes the error term $e[n]$, which is used to update the coefficients of the FIR filter. Figure L7–1 shows this process.

108 C6x-Based DSP

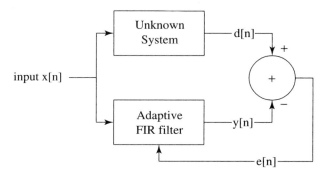

FIGURE L7-1 Adaptive filtering.

The error term calculated from the difference of the outputs of the two systems is used to update each coefficient of the FIR filter in the following manner (least-mean-square (LMS) algorithm [17]):

$$h_n[k] = h_{n-1}[k] + \Delta * e[n] * x[n - k],$$

where h's denote the unit sample response or FIR filter coefficients.

The output $y[n]$ is required to approach $d[n]$. The term Δ indicates the step size for the gradient descent method. A small step size will ensure convergence, but results in a slow adaptation rate. A large step size, though faster, may lead to skipping over the solution. For our case, we are limited by the smallest number representable by Q-15 format (i.e., 0x0001, which is in the range of $1e^{-4}$).

L7.1 Design of IIR Filter

A seventh-order bandpass IIR filter is used to act as the unknown system. The adaptive FIR should adapt to the response of the IIR. Using a sampling frequency of 8 KHz, the specifications of the IIR filter may be stated as having a passband from $\pi/3$ to $2\pi/3$ (radians), with a stopband attenuation of 20 dB. Using the yulewalk function of MATLAB [15], the design of this filter can be easily achieved. The following MATLAB code may be used to obtain the coefficients of the filter. It can be verified that the filter is working by deploying a simple composite signal. Using the filter function of MATLAB, the design may be verified by observing that the frequency components of the composite signal falling in the stopband are removed. (See Figure L7-2 and Table L7-1.)

```
Nc=7;
f=[0 0.32 0.33 0.66 0.67 1];
m=[0 0 1 1 0 0];
[B,A]=yulewalk(Nc,f,m);
freqz(B,A);

%Create A sample signal
Fs=8000;
Ts=1/Fs;
Ns=128;
t=[0:Ts:Ts*(Ns-1)];
f1=750;
f2=2000;%The one to keep
f3=3000;

x1=sin(2*pi*f1*t);
x2=sin(2*pi*f2*t);
x3=sin(2*pi*f3*t);

x=x1+x2+x3;
%Filter it
y=filter(B,A,x);
```

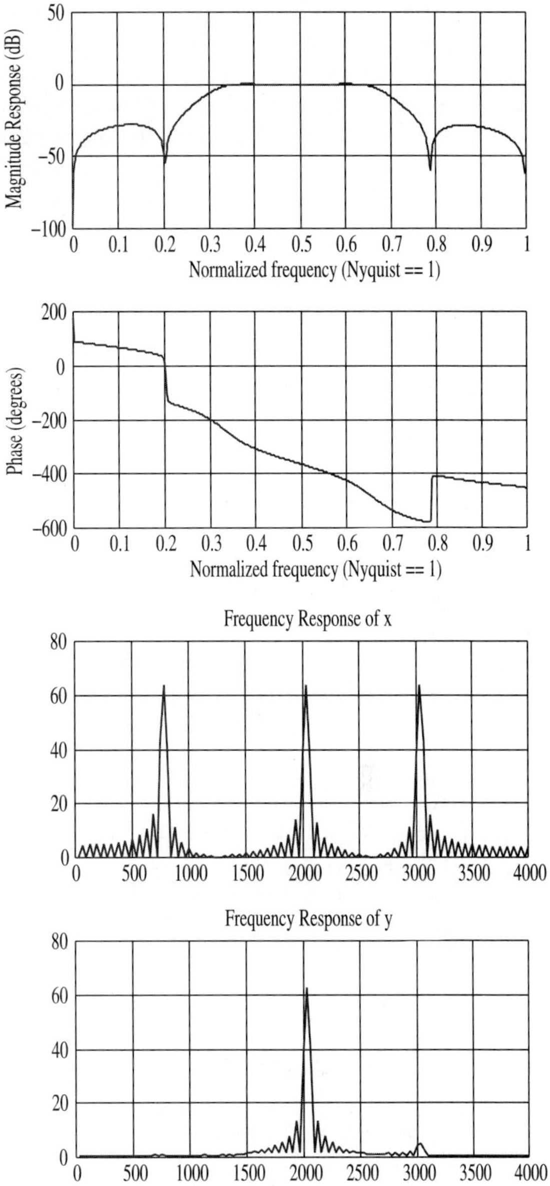

TABLE L7–1 IIR filter coefficients.

A's	B's
1.0000	0.1191
0.0179	0.0123
0.9409	−0.1813
0.0104	−0.0251
0.6601	0.1815
0.0342	0.0307
0.1129	−0.1194
0.0058	−0.0178

Note: Do not confuse A&B coefficients with the CPU registers!

FIGURE L7–2 Filter response.

L7.2 IIR Filter Implementation

The implementation of the IIR filter on the C62 EVM is initially done in C. Two arrays are required: one for the input samples $x[n]$ and the other for the output samples $y[n]$. Given that the filter is of the order seven, an input array of size eight and an output array of size seven are considered. The arrays are used to simulate a circular buffer, since, in C, this property of the hardware cannot be accessed. As a new sample comes in, all elements in the input array are shifted down by one, losing the last element. In this manner, the last eight samples are always kept. The input array is used to calculate the resulting output, and then the output is used to modify the output array using the same approach. A simple implementation of this scheme is shown in the following code block, which is a modification of the sampling program in Lab 3.

```
interrupt void serialPortRcvISR (void)
{
   int temp,n,ASUM,BSUM;
   short input,IIR_OUT,output;

        temp = MCBSP_READ(0);
        input = temp >> S;

        for(n=7;n>0;n--)         //Input buffer
               IIRwindow[n] = IIRwindow[n-1];

        IIRwindow[0] = input;

        BSUM = 0;

        for(n=0;n<=7;n++)
        {                                                //Multiplication of Q-15 with Q-15
               BSUM += ((BS[n] * IIRwindow[n]) << 1);    //Results in Q-30.  Shift by one to
        }                                                //Eliminate Sign Extension bit

        ASUM = 0;
        for(n=0;n<=6;n++)
        {
               ASUM += ((AS[n] * y_prev[n]) << 1);
        }

        IIR_OUT = (BSUM - ASUM) >> 16;
        for(n=6;n>0;n--)         //Output buffer
               y_prev[n] = y_prev[n-1];

        y_prev[0] = IIR_OUT;

        MCBSP_WRITE(0, IIR_OUT << S);
}
```

Note that the arrays IIRwindow and y_prev are global and that all data is represented in Q-15 format. The coefficients A's and B's have been converted to Q-15 format by multiplying them with 0x7FFF. The scaling variable S can be used to correct for possible overflow, but here using the usual shift of 16 is adequate. By running this program while connecting a function generator and an oscilloscope to the line-in and line-out ports of the EVM, the functionality of the IIR filter can be easily verified.

L7.3 Adaptive FIR Filter

A 32-coefficient FIR filter is used to adapt to the output of the IIR filter. To do this in C, an additional buffer of length 32 is needed. Two arrays are required: one for the input buffer and the other for the coefficients of the FIR filter. Initially all data in both arrays are zero. The order of processing is as follows. First, the signal is filtered by computing the dot product of the coefficients and the input buffer. Using the output from the filtering operation, the error term between the IIR and the FIR output is calculated. The coefficients of the FIR filter are then updated accordingly. The following piece of code illustrates how this is achieved in C:

```
//Seperate Circular buffer for FIR
for(n=31;n>0;n-)
        FIRwindow[n] = FIRwindow[n-1];

FIRwindow[0] = input;

//Perform Filtering with current coefficients
temp = 0;
for(n=0;n<32;n++)
{
        temp += ((h[n]*FIRwindow[n]) << 1);
}

y = temp >> 16;

//Calculate Error Term

e = IIR_OUT - y;

//Update Coefficients

stemp = (DELTA*e)>>15;

for(n=0;n<32;n++)
{
        stemp2 = (stemp*FIRwindow[n])>>15;
        h[n] = h[n] + stemp2;
}
```

By adding this piece of code to the previous IIR code, we can now use the FIR filter to adapt to the output of the IIR filter. By using the function generator and the oscilloscope, we can observe the adaptation process as we cycle through different frequencies.

It is worth mentioning a point about the step size Δ here. In a floating-point processor, Δ is usually chosen to be in the range of $1e^{-7}$. Note that the precision on the C62 does not allow for such a small number. We can at most use 0x0001, which is in the range of

$1/(2^{15}) \approx 0.0000305$. When a multiplication is done with this number, any positive number will be defaulted to zero and any negative number to -1. This is due to the nature of multiplication of Q-15 format numbers where the product is right shifted by 15. However, the contribution of negative numbers to the coefficients is sufficient for the LMS algorithm to converge. Using a larger Δ, for this application, results in faster adaptation, but convergence would not be guaranteed. Satisfactory results can be observed with Δ in the range of 0x0100 to 0x0001.

The main point here is to have very small adjustments made to the coefficients. The step size Δ actually serves this purpose. The error term and the input are multiplied and scaled by Δ before being added to the previous coefficients. The ratio of the magnitudes of the coefficients to the modification term should be rather high, so that the modifications do not cause skipping over the solution.

The reason for implementing the LMS algorithm in assembly is to make use of the circular buffering capability of the C6x. Of the 32 registers on the C6x, 8 of them can perform circular addressing. These registers are A4 through A7 and B4 through B7. Since linear addressing is the default mode of addressing, each of these registers must be specified as circular using the AMR. The lower 16 bits of the AMR are used to select the mode for each of the 8 registers. The upper 10 bits (6 are reserved) are used to set the length of the circular buffer. Buffer size is determined by $2^{(N+1)}$ bytes, where N is the value appearing in the block size fields of the AMR.

Since we are using both C and assembly, we have to initialize the circular buffer only when we enter the assembly part of the program. When the assembly code is executed, the register that is used in circular mode will ensure that a certain location in memory always contains the newest sample. As the code completes and returns to the calling C program, the location of the pointer in the buffer must be saved, and the AMR register must be returned to linear mode, since leaving it in circular mode may disrupt the flow of the C program.

To do such a task, a section of memory must be set aside for the buffer and the coefficients and not used by the compiler. A simple way to do this is to reserve 16 bytes for the coefficients, 16 for the buffer, and 1 for the pointer location. Since the data and coefficients are short formatted here, two 16 bytes are used to provide 32 locations. The following memory representation is used for this purpose:

8000_0000	16 Bytes
	Circular Buffer
8000_0040	16 Bytes
	Coefficients
8000_0080	1 Byte, Pointer

The command file must also be modified. A simple assembly directive file is needed to initialize the memory locations with zeros. The following command file defines a new memory section called MMEM in the internal data memory and uses it for the section ".mydata". The file *initmem.asm*, shown next, is used to initialize the memory locations with zeros and set the pointer to the first free location, which is 8000_0000.

```
MEMORY
{
  INT_PROG_MEM     (RX)    : origin = 0x00000000  length = 0x00010000
  SBSRAM_PROG_MEM  (RX)    : origin = 0x00400000  length = 0x00014000
  SBSRAM_DATA_MEM  (RW)    : origin = 0x00414000  length = 0x0002C000
  SDRAM0_DATA_MEM  (RW)    : origin = 0x02000000  length = 0x00400000
  SDRAM1_DATA_MEM  (RW)    : origin = 0x03000000  length = 0x00400000
  INT_DATA_MEM     (RW)    : origin = 0x80000100  length = 0x0000FF00
  MMEM                     : origin = 0x80000000  length = 0x00000100
}

SECTIONS
{
  .vec:       load = 0x00000000
  .text:      load = SBSRAM_PROG_MEM
  .const:     load = INT_DATA_MEM
  .bss:       load = INT_DATA_MEM
  .data:      load = INT_DATA_MEM
  .cinit      load = INT_DATA_MEM
  .pinit      load = INT_DATA_MEM
  .stack      load = INT_DATA_MEM
  .far        load = INT_DATA_MEM
  .sysmem     load = SDRAM0_DATA_MEM
  .cio        load = INT_DATA_MEM
  .int        load = INT_PROG_MEM
  .mydata     load = MMEM
  sbsbuf      load = SBSRAM_DATA_MEM
        { _SbsramDataAddr = .; _SbsramDataSize = 0x0002C000; }
}
```

initmem.asm

```
.sect ".mydata"
      .short   0,0,0,0,0,0,0,0,0,0,0,0,0,0,0,0,0,0,0,0,0,0,0,0,0,0,0,0,0,0,0,0
      .short   0,0,0,0,0,0,0,0,0,0,0,0,0,0,0,0,0,0,0,0,0,0,0,0,0,0,0,0,0,0,0,0
      .field   0x80000000,32
```

With the command file and the small assembly code just shown, it is ensured that the 33 bytes of space starting at 8000_0000 will not be used for anything other than the adaptive filter. Now, as mentioned before, the circular buffer must be initialized upon entering the assembly part. To do this, it is necessary to modify the AMR. Since a buffer of length 32 is needed, it is required to have 4 in the block fields (block size = $2^{(4+1)}$ = 32). Using register A5 as the circular buffer pointer, the value to set the AMR hence becomes 0x00040005. Entering the assembly function, the last pointer location is read from 8000_0080. The last free location of the buffer is saved to the same location upon exit. The following code shows how this is achieved. Note that here a dummy load is performed to increment the pointer so that it points to the last element (next free location). This is performed because only a load or store operation increments the pointer in a circular fashion.

```
;Initialize the Circular buffer for the FIR filter
MVK             .S2     0x0004,B10              ;A5 is selected as circular
MVKLH           .S2     0x0005,B10   ;2^(4+1)=32
MVC             .S2     B10,AMR

;Load the pointer to A0
;Assume that the current location of the circular buffer
;is pointed to

MVK             .S1     0x0080,A0
MVKLH           .S1     0x8000,A0;A0=0x80000080 (has last pointer)

LDW             .D1     *A0,A5
NOP             4                               ;A5 now points to the first free
                                                ;Location of the Circular buffer
;Load the current sample to the Circular buffer
STH             .D1     A4,*A5   ;A4 has sample passed from Calling

//FIR FILTERING HERE

;Now save the Last location of A0 to memory

MVK             .S1     0x0080,A0
MVKLH           .S1     0x8000,A0       ;This address has the pointer to x

LDH             .D1 *A5++,A13    ;Dummy Load
STW             .D1 A5,*A0       ;Saved last
;Restore Linear Addressing
MVK             .S2     0x0000,B10
MVKLH           .S2     0x0004,B10
MVC             .S2     B10,AMR

;return the result y

MV              .S1     A9,A4
B               .S2     B3
NOP             5
```

The adaptive FIR algorithm resides in the section of the foregoing code labeled FIR HERE. A simple approach to the adaptive filtering process, which involves calculating the error term and coefficient updating, is to have two separate loops. The first loop simply calculates the dot product of the coefficients and the samples. The error term is calculated and used in the second loop to update the coefficients based on the input samples. Notice that, in this way, a circular buffer is not used for the coefficients, since they do not change in a time-windowed manner, as do the input samples.

We now have two versions of the adaptive filter. One is written completely in C, and the other is a mix of C and assembly. The C code has to be modified to run from the internal memory space. This can be accomplished by compiling the interrupt in a separate file and keeping the resulting assembly file. The section directive of the file has to be modified to reflect the internal memory space, and the large memory model should be used during compilation. The large memory model is required to compensate for the large offset existing between the program in the internal memory and the program in the external memory. To use the large memory model, the -m3 option needs to be included in the command line. Using this approach, we let the entire interrupt run from the internal memory, allowing a faster execution time.

The assembly version of the program also needs to be rewritten in linear assembly to allow the compiler to create software-pipelined code. All that needs to be done is to change the register names to symbolic variable names and add several precompiler directives such as .trip and .proc. Table L7–2 gives a summary of the timing cycles for different versions of the adaptive filtering program.

TABLE L7–2   Timing cycles for the adaptive filter program

C program in external memory	9455
C program in internal memory	3720
Non-optimized Assembly in external memory	3577
Linear assembly in external memory	2850
Linear assembly in internal memory	1153

The assembly version of the program is also configured to run in the internal memory space, where the number of cycles is noticeably reduced. The main reason for running the C code from the internal memory space is that when it runs from the external memory, it may miss samples, since it runs too slow. A note to remember is that the large memory model has to be used, since the main program that calls the interrupt from the internal memory is in the external memory space. As a result, a far pointer is required for the large offset.

The complete listing of the programs for this lab appears in Appendix C.

CHAPTER 9

Application Examples

This final chapter provides two application examples to further illustrate different implementations generating different timing results. The first example involves a wavelet-based denoising algorithm and the second one an efficient algorithm for wavelet reconstruction. These examples can be treated as two additional lab exercises; the codes for both examples are placed on the attached CD. Additional application examples are available at the TI C6000 Web site *http://www.ti.com/sc/docs/products/dsp/c6000/index.htm*.

9.1 WAVELET-BASED DENOISING

Wavelet transform is a recently developed signal representation scheme that provides both time and frequency localization, as compared with the Fourier transform, which is not localized in time. For theoretical details on wavelet transform, refer to references on this subject. (For example, see [18].) In wavelet analysis, low- and high-frequency components are called approximations and details, respectively. Approximations correspond to high-scale, low-frequency components, while details correspond to low-scale, high-frequency components of the signal. An efficient and practical wavelet filtering algorithm was developed by Mallat [19]. In this algorithm, a signal is decomposed into its approximations and details components by using two complementary filters, one a lowpass and the other a highpass, generating two new signals. Since twice as much data are produced, a downsampling mechanism (meaning throwing away every second sample) is adopted to keep the signal length unchanged. The decomposition process is iterated with successive decompositions of approximations.

The denoising algorithm implemented here is illustrated in Figure 9–1. First, the signal is decomposed as shown in this figure to generate the wavelet coefficients. Then,

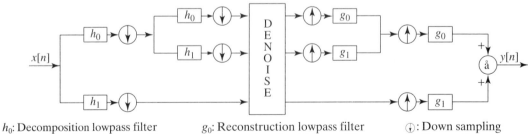

h_0: Decomposition lowpass filter
h_1: Decomposition highpass filter
g_0: Reconstruction lowpass filter
g_1: Reconstruction highpass filter
↓: Down sampling
↑: Upsampling

FIGURE 9–1   Diagram of wavelet denoising algorithm.

the details coefficients that are below a specified threshold level are removed. The threshold chosen is sqrt(2 * log(decomposition level)), as stated in [20]. The signal is then reconstructed based on the original approximation coefficients and the thresholded details coefficients. Here the coefficients corresponding to the 16th-order Daubechies wavelet are used for denoising. The MATLAB code appearing on the CD helps one to understand the decomposition, thresholding, and reconstruction as carried out in the corresponding C program.

Table 9–1 includes the timing results for the C program using different levels of optimization. The linear assembly entry corresponds to the situation when using the linear assembly version of the FIR functions.

TABLE 9–1 Timing cycles for wavelet denoising algorithm

Code Type		Cycles of ISR
C		6,863,352
Optimizations	-o0	4,003,715
	-o1	2,877,645
	-o2	945,122
	-o3	945,122
Linear Assembly		733,062

9.2 EFFICIENT WAVELET RECONSTRUCTION

Figure 9–2 shows the structure for dyadic wavelet decomposition and reconstruction. The decomposition and the reconstruction are normally done by using the convolution operator.

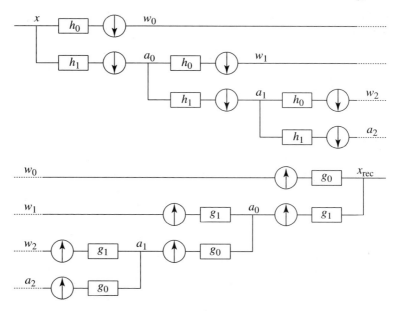

FIGURE 9–2 Wavelet decomposition and reconstruction.

In the algorithm implemented here and described in [21], the reconstruction is done by overlapping outer products saving one-half of the usual number of multiplications. In addition, the decomposition is performed using the correlation instead of the usual convolution operator. As a result, the reconstruction filters are used directly for decomposition, without the need to use their flipped versions. In this implementation, the four-coefficient Daubechies wavelet is used because of its good time–frequency localization properties. In what follows, a description of the implemented decomposition and reconstruction codes is provided. These codes appear on the attached CD.

An input buffer of 2048 samples is decomposed and reconstructed up to the seventh decomposition level. The four-coefficient Daubechies wavelet used requires the allocation of several buffers. Also, some extra memory locations are allocated at the beginning and end of each buffer for the purpose of simplifying the computation of convolution at these borders.

The buffers corresponding to the scaling function are named a_i's, and the ones corresponding to the wavelet are named w_i's. The number of extra locations is given by the length of the wavelet filter minus one. Hence, the array lengths are determined as follows:

$$\text{samples: } 3 + 2048 + 3 = 2054$$

$$w_0: \text{round}[(2048 + 3)/2] = 1026$$

$$a_0: 3 + 1026 + 3 = 1032$$

$$w_1: \text{round}[(1026 + 3)/2] = 515$$

$$a_1: 3 + 515 + 3 = 521$$

$$w_2: \text{round}[(515 + 3)/2] = 259$$

$$a_2: 3 + 259 + 3 = 265$$

$$w_3: \text{round}[(259 + 3)/2] = 131$$

$$a_3: 3 + 131 + 3 = 137$$

$$w_4: \text{round}[(131 + 3)/2] = 67$$

$$a_4: 3 + 67 + 3 = 73$$

$$w_5: \text{round}[(67 + 3)/2] = 35$$

$$a_5: 3 + 35 + 3 = 41$$

$$w_6: \text{round}[(35 + 3)/2] = 19$$

$$a_6: 19$$

The provided wave function cleans the beginning and the end of each buffer corresponding to the scaling function. Then, the wavelet decomposition is performed by calling the decomp routine seven times. After the decomposition is done, all the arrays corresponding to the scaling function (a_0 through a_5), except for the a_6 array and the samples array, are cleaned. This is needed because the elements of those arrays are used as accumulators during the reconstruction phase. Finally, the `recons` routine is called seven times to reconstruct the input signal.

The decomposition procedure is performed by the `decomp` routine. Instead of using the convolution operator, the decomposition is done using the correlation operator. This routine requires six arguments in the following order:

1. *signal*: A pointer pointing to the signal to be decomposed.
2. a_0: A pointer pointing to the array that corresponds to the scaling function part of the input signal.
3. w_0: A pointer pointing to the array that corresponds to the wavelet part of the input signal.
4. *length*: This is the number of samples in the signal array plus three. For example, when the decomposition routine is called for the first time, *length* $= 2048 + 3$, the second time *length* $=$ round $[(2048 + 3)/2] + 3 = 1029$, and so on.
5. h_0: The filter corresponding to the scaling function.
6. h_1: The filter corresponding to the wavelet.

Note that h_0 and h_1 are the same as g_0 and g_1, respectively. They are not flipped versions of g_0 and g_1, as is usually done, since correlation instead of convolution is employed to perform the decomposition. The result of the correlation is downsampled by two and all the fixed-point arithmetic is handled by using the Q-15 number representation.

The reconstruction procedure is then performed by the `recons` routine. This routine uses outer product operations rather than the conventional upsampling and convolution operations. From an implementation point of view, this is more efficient, since, in this manner, the upsampling and multiplication by zero are avoided. Consider upsampling by two a w array $w = [w_1\ w_2\ w_3\ w_4\ w_5\ldots]$ followed by convolution, denoted by $\otimes$, with the filter $g_0 = [a\ b\ c\ d]$. This is equivalent to the right-hand side of the following equation:

$$\begin{aligned} w \otimes g_0 = &\ aw_1 \quad bw_1 \quad cw_1 \quad dw_1 \\ + & \qquad\qquad aw_2 \quad bw_2 \quad cw_2 \quad dw_2 \\ + & \qquad\qquad\qquad\qquad aw_3 \quad bw_3 \quad cw_3 \quad dw_3 \\ + & \qquad\qquad\qquad\qquad\qquad\qquad \ldots\ldots\ldots\ldots \end{aligned}$$

After summing the contents of each column on the right-hand side, it is easy to see that the left-hand side sequence and the right-hand side sequence are the same. Notice that the multiplications by zero in the conventional convolution approach on the left-hand side are saved. This routine also requires the same six arguments as `decomp`.

It should be noted that this algorithm is done in a triple buffering fashion. The interrupt service routine is used to allocate a frame of input samples to the input buffer. At the same frame time, a previously collected frame of samples is decomposed/reconstructed, and a previously reconstructed frame of samples residing in the output buffer is sent to the codec.

Table 9–2 shows the timing results for this more efficient wavelet reconstruction algorithm. It is interesting to note that, in this case, the fully optimized C code performs better than the linear assembly version. This is primarily due to the fact that the compiler optimizer unrolls the inner loop in the decomposition/reconstruction routine, whereas the assembler optimizer does not do so for this case.

TABLE 9–2 Timing cycles for wavelet reconstruction algorithm

Code Type	Cycles
C (-g –O3)	158567
C (-g)	852785
Linear assembly	394421

APPENDIX A

FIR Filter Program Listing for Lab 5

Main program for FIR filter: *evm_fir.c*

```c
#include <stdlib.h>
#include <stdio.h>
#include <string.h>
#include <common.h>
#include <mcbspdrv.h>
#include <intr.h>
#include <board.h>
#include <codec.h>
#include <mcbsp.h>
#include <mathf.h>

#define PRINT_DBG 1

/*------------------------------------*/
/* GLOBALS
*/
/*------------------------------------*/

// 11 tap fir filter coeffs
short b[] =
{0x6DF,0x0,0xF447,0xFFFD,0x2813,0x3FFC,0x2813,0xFFFD,0xF447,0x0,0x6DF};
int samples[]={0,0,0,0,0,0,0,0,0,0,0};

/*------------------------------------*/
/* FILE LOCAL (STATIC) PROTOTYPES
*/
/*------------------------------------*/
void hookint(void);
interrupt void serialPortRcvISR (void);

int FIR_filter(int* samples, short* coeffs);  //FIR in C
extern fir_simple();        //unoptimized FIR in assembly
extern fir_filled();        //optimized FIR in assembly
extern fir_la();            //FIR in linear assembly
```

```c
/*------------------------------------------*/
/* Main
*/
/*------------------------------------------*/
main()
{
        Mcbsp_dev    dev;
        Mcbsp_config mcbspConfig;
        int i,sampleRate,status,actualrate;

        /*********************************************/
        /* Initialize EVM                            */
        /*********************************************/

        evm_init();

        /*************************************************/
        /* Open MCBSP for subsequent Examples            */
        /*************************************************/

        mcbsp_drv_init();

        dev = mcbsp_open(0);

        if (dev == NULL)
        {
                printf("Error opening MCBSP 0    \n ");
                return(ERROR);
        }
        /*********************************************/
        /* configure McBSP                           */
        /*********************************************/

        memset(&mcbspConfig,0,sizeof(mcbspConfig));

        mcbspConfig.loopback          = FALSE;

        mcbspConfig.tx.update         = TRUE;
        mcbspConfig.tx.clock_mode     = CLK_MODE_EXT;
        mcbspConfig.tx.frame_length1  = 0;
        mcbspConfig.tx.word_length1   = WORD_LENGTH_32;

        mcbspConfig.rx.update         = TRUE;
        mcbspConfig.rx.clock_mode     = CLK_MODE_EXT;
        mcbspConfig.rx.frame_length1  = 0;
        mcbspConfig.rx.word_length1   = WORD_LENGTH_32;

        mcbsp_config(dev,&mcbspConfig);

        MCBSP_ENABLE(0, MCBSP_BOTH);
```

Appendix A   FIR Filter Program Listing for Lab 5

```c
/**********************************************/
/* configure CODEC                             */
/**********************************************/
codec_init();

/* A/D 0.0 dB gain, turn off 20dB mic gain, sel (L/R)LINE input  */
codec_adc_control(LEFT,0.0,FALSE,LINE_SEL);
codec_adc_control(RIGHT,0.0,FALSE,LINE_SEL);

/* mute (L/R)LINE input to mixer              */
codec_line_in_control(LEFT,MIN_AUX_LINE_GAIN,TRUE);
codec_line_in_control(RIGHT,MIN_AUX_LINE_GAIN,TRUE);


/* D/A 0.0 dB atten, do not mute DAC outputs  */
codec_dac_control(LEFT, 0.0, FALSE);
codec_dac_control(RIGHT, 0.0, FALSE);


sampleRate = 8000;

actualrate = codec_change_sample_rate(sampleRate, TRUE);

codec_interrupt_enable();

hookint();

/********************************************************************/
/* Main Loop: wait for Interrupt                                    */
/********************************************************************/

for(i=0;i<=N;i++)
        samples[i]=0;

while (1)
{
}

/* mute DAC outputs                           */
codec_dac_control(LEFT, 0.0, TRUE);
codec_dac_control(RIGHT, 0.0, TRUE);

mcbsp_close(dev);

return(OK);
}
```

124   C6x-Based DSP

```
/*------------------------------------*/
/* FUNCTIONS                          */
/*------------------------------------*/

interrupt void serialPortRcvISR (void)
{
        int i,temp,result;

        result = 0;

        temp = MCBSP_READ(0);

        //update array samples
        for(i=N-1;i>=0;i--)
                samples[i+1]=samples[i];

        samples[0] = temp;

        //Filtering, use only one of the functions
        result = FIR_filter(samples,b);
        //result = fir_simple(samples,b,11);
        //result = fir_filled(samples,b,11);
        //result = fir_la(samples,b,11);

        MCBSP_WRITE(0, result);
}

void hookint()
{
        intr_init();
        intr_map(CPU_INT15, ISN_RINT0);
        intr_hook(serialPortRcvISR, CPU_INT15);

        INTR_ENABLE(15);
        INTR_GLOBAL_ENABLE();

        return;
}

int FIR_filter(int* samples, short* coeffs)
{
        int i,result=0;
        //Filtering
        for(i=0;i<=N;i++)
                result += ( _mpyhl(samples[i],b[i]) ) << 1;

        return result;
}
```

Appendix A FIR Filter Program Listing for Lab 5

Assembly file: *fir_simple.asm*

```
        .global  _fir_simple

        .sect    ".fir_simple"

_fir_simple:
        MV      .S1     A6,A2           ;Count register
        ZERO    .S1     A9              ;Sum register

loop:   LDW     .D1     *A4++,A7        ;Load data from samples
        LDH     .D2     *B4++,B7        ;Load data from Coefficients
        NOP     4
        MPYHL   .M1x    A7,B7,A8        ;A7 is 32 bit sample, B7 is Q-15
                                         representation coefficient
        NOP
        SHL             A8,1,A8         ;Eliminate sign extension bit
        ADD     .S1     A8,A9,A9        ;Accumulate result
[A2]    SUB     .S1     A2,1,A2         ;Decrement counter
[A2]    B       .S1     loop
        NOP     5

        MV      .S1     A9,A4           ;Move result to return register
        B       .S2     B3              ;Branch back to calling address
        NOP     5
```

Delay slot filled assembly file: *fir_filled.asm*

```
        .global  _fir_filled

        .sect    ".fir_filled"

_fir_filled:
        MV      .S1     A6,A2           ;Count register
        ZERO    .S1     A9              ;Sum register

loop:   LDW     .D1     *A4++,A7        ;Load data from samples
||      LDH     .D2     *B4++,B7        ;Load data from Coefficients
        NOP
[A2]    SUB     .S1     A2,1,A2         ;Decrement counter
[A2]    B       .S1     loop
        NOP
        MPYHL   .M1x    A7,B7,A8        ;A7 is 32 bit sample, B7 is Q-15
                                         representation coefficient
```

```
        NOP
        SHL     A8,1,A8                 ;Eliminate sign extension bit
        ADD     .S1    A8,A9,A9         ;Accumulate result

        MV      .S1    A9,A4            ;Move result to return register
        B       .S2    B3               ;Branch back to calling address
        NOP     5
```

FIR in linear assembly: *fir_la.sa*

```
        .global _fir_la

        .sect   ".fir_la"

_fir_la:    .proc   a4,b4,a6,b3
            .reg    p_m,m,p_n,n,prod,sum,cnt

            mv      a4, p_m
            mv      b4, p_n
            mv      a6, cnt
            zero    sum

loopLA:     .trip   11

        ldw     *p_m++, m
        ldh     *p_n++, n
        mpyhl   m,n,prod
        shl     prod,1,prod
        add     prod,sum,sum

[cnt]   sub     cnt,1,cnt
[cnt]   b       loopLA

        mv      sum,a4
        .endproc a4,b3

        B       B3
        NOP     5
```

Linker command file for FIR lab: *link.cmd*

```
MEMORY
{
   INT_PROG_MEM    (RX)    : origin = 0x00000000  length = 0x00010000
   SBSRAM_PROG_MEM (RX)    : origin = 0x00400000  length = 0x00014000
   SBSRAM_DATA_MEM (RW)    : origin = 0x00414000  length = 0x0002C000
   SDRAM0_DATA_MEM (RW)    : origin = 0x02000000  length = 0x00400000
   SDRAM1_DATA_MEM (RW)    : origin = 0x03000000  length = 0x00400000
   INT_DATA_MEM    (RW)    : origin = 0x80000000  length = 0x00010000
}
```

Appendix A   FIR Filter Program Listing for Lab 5

```
SECTIONS
{
    .vec:           load = 0x00000000
    .text:          load = SBSRAM_PROG_MEM
    .const:         load = INT_DATA_MEM
    .bss:           load = INT_DATA_MEM
    .data:          load = INT_DATA_MEM
    .cinit          load = INT_DATA_MEM
    .pinit          load = INT_DATA_MEM
    .stack          load = INT_DATA_MEM
    .far            load = INT_DATA_MEM
    .sysmem         load = SDRAM0_DATA_MEM
    .cio            load = INT_DATA_MEM
    .fir_filled:    load = INT_PROG_MEM
    .fir_simple:    load = INT_PROG_MEM
    .fir_la:        load = INT_PROG_MEM
    sbsbuf          load = SBSRAM_DATA_MEM
                    { _SbsramDataAddr = .; _SbsramDataSize = 0x0002C000; }
}
```

Build commands: *buildfir.bat*

```
asm6x fir_simple.asm
asm6x fir_filled.asm
cl6x fir_la.sa
cl6x -op0 -ml3 -c -g -i. -iC:\Evm6x\Dsp\Include -iC:\C6xTools\include  evm_fir.c
lnk6x -m fir.map -c -o fir.out evm_fir.obj fir_simple.obj fir_filled.obj fir_la.obj
      c:\Evm6x\Dsp\Lib\Drivers\Drv6x.lib C:\Evm6x\Dsp\Lib\DevLib\Dev6x.lib
        C:\C6xTools\lib\rts6201.lib
      link.cmd
```

Main program for FIR filter (C67 version): *fir_evm67.c*

```
#include <stdlib.h>
#include <string.h>
#include <common.h>
#include <mcbspdrv.h>
#include <intr.h>
#include <board.h>
#include <codec.h>
#include <mcbsp.h>
```

```c
/*---------------------------------------------*/
/* FILE LOCAL (STATIC) PROTOTYPES
*/
/*---------------------------------------------*/
void hookint(void);
interrupt void serialPortRcvISR (void);
float dotp1(const float a[], const float b[]);
float coef[11]={ 0.0537, 0.0 , -0.0916, -0.0001, 0.3131, 0.4999,
        0.3131, -0.0001, -0.0916, 0.0 ,0.0537};
float x[11] = {0.0,0.0,0.0,0.0,0.0,0.0,0.0,0.0,0.0,0.0,0.0};
main()
{
        Mcbsp_dev dev;
        Mcbsp_config   mcbspConfig;
        int sampleRate,status,actualrate;
        evm_init();
        mcbsp_drv_init();
        dev = mcbsp_open(0);
        /***********************************************/
        /* configure McBSP                             */
        /***********************************************/
        memset(&mcbspConfig,0,sizeof(mcbspConfig));
        mcbspConfig.loopback           = FALSE;
        mcbspConfig.tx.update          = TRUE;
        mcbspConfig.tx.clock_mode      = CLK_MODE_EXT;
        mcbspConfig.tx.frame_length1   = 0;
        mcbspConfig.tx.word_length1    = WORD_LENGTH_32;
        mcbspConfig.rx.update          = TRUE;
        mcbspConfig.rx.clock_mode      = CLK_MODE_EXT;
        mcbspConfig.rx.frame_length1   = 0;
        mcbspConfig.rx.word_length1    = WORD_LENGTH_32;
        mcbsp_config(dev,&mcbspConfig);
        MCBSP_ENABLE(0, MCBSP_BOTH);
        codec_init();
        /* A/D 0.0 dB gain, turn on 20dB mic gain, sel (L/R)LINE input*/
        codec_adc_control(LEFT,0.0,FALSE,LINE_SEL);
        codec_adc_control(RIGHT,0.0,FALSE,LINE_SEL);
        /* mute (L/R)LINE input to mixer                    */
        codec_line_in_control(LEFT,MIN_AUX_LINE_GAIN,TRUE);
        codec_line_in_control(RIGHT,MIN_AUX_LINE_GAIN,TRUE);
        /* D/A 0.0 dB atten, do not mute DAC outputs        */
        codec_dac_control(LEFT, 0.0, FALSE);
        codec_dac_control(RIGHT, 0.0, FALSE);
        sampleRate = 8000;
        actualrate = codec_change_sample_rate(sampleRate, TRUE);
        codec_interrupt_enable();
        hookint();
```

```
            /****************************************************************/
            /* Main Loop: wait for Interrupt                                */
            /****************************************************************/

            while (1)
            {
            }
            return(OK);
}

void hookint()
{
            intr_init();
            intr_map(CPU_INT15, ISN_RINT0);
            intr_hook(serialPortRcvISR, CPU_INT15);

            INTR_ENABLE(15);
            INTR_GLOBAL_ENABLE();

            return;
}

float dotp1(const float a[], const float b[])
{
            int i;
            float sum = 0.0;

            for(i=0; i<11;i++)
                    sum += a[i] * b[i];

            return sum;
}
interrupt void serialPortRcvISR (void)
{
            int i,sample;
            float temp, sum;
            sample = MCBSP_READ(0);

            temp = (float)sample;

            for(i=10;i>=0;i--)
                    x[i]=x[i-1];

            x[0]=temp;

            sum = dotp1(coef, x);

            MCBSP_WRITE(0, sum);
}
```

Build commands for FIR (C67 version): *build67.bat*

```
asm6x -mv6700 dotp2.asm
cl6x -c -q -k -ss -g -mv6700 -i. -iC:\ti\Evm6x01\Dsp\Include -iC:\C6xTools\include  adapt.c
lnk6x -c -o fir.out dotp2.obj adapt.obj C:\ti\Evm6x01\Dsp\Lib\Drivers\Drv6x.lib
         c:\ti\Evm6x01\Dsp\Lib\DevLib\Dev6x.lib C:\C6xTools\lib\rts6701.lib link.cmd
```

APPENDIX B

FFT Program Listing for Lab 6

Twiddle factors for FFT: *twiddleI.h* and *twiddleR.h*

```
float TI[64] = {0.0000,0.0980,0.1951,0.2903,0.3827,0.4714,0.5556,0.6344,
0.7071,0.7730,0.8315,0.8819,0.9239,0.9569,0.9808,0.9952,0.9999,0.9952,
0.9808,0.9569,0.9239,0.8819,0.8315,0.7730,0.7071,0.6344,0.5556,
0.4714,0.3827,0.2903,0.1951,0.0980,0.0000,-0.0980,-0.1951,-0.2903,
-0.3827,-0.4714,-0.5556,-0.6344,-0.7071,-0.7730,-0.8315,-0.8819,
-0.9239,-0.9569,-0.9808,-0.9952,-1.0000,-0.9952,-0.9808,-0.9569,
-0.9239,-0.8819,-0.8315,-0.7730,-0.7071,-0.6344,-0.5556,-0.4714,
-0.3827,-0.2903,-0.1951,-0.0980};

float TR[64] ={0.9999,0.9952,0.9808,0.9569,0.9239,0.8819,0.8315,0.7730,
0.7071,0.6344,0.5556,0.4714,0.3827,0.2903,0.1951,0.0980,0.0000,-0.0980,
-0.1951,-0.2903,-0.3827,-0.4714,-0.5556,-0.6344,-0.7071,-0.7730,
-0.8315,-0.8819,-0.9239,-0.9569,-0.9808,-0.9952,-1.0000,-0.9952,
-0.9808,-0.9569,-0.9239,-0.8819,-0.8315,-0.7730,-0.7071,-0.6344,
-0.5556,-0.4714,-0.3827,-0.2903,-0.1951,-0.0980,0.0000,0.0980,0.1951,
0.2903,0.3827,0.4714,0.5556,0.6344,0.7071,0.7730,0.8315,0.8819,0.9239,
0.9569,0.9808,0.9952};
```

Linker command file for FFT: *LinkFFT.cmd*

```
MEMORY
{
  INT_PROG_MEM    (RX)   : origin = 0x00000000 length = 0x00010000
  SBSRAM_PROG_MEM (RX)   : origin = 0x00400000 length = 0x00014000
  SBSRAM_DATA_MEM (RW)   : origin = 0x00414000 length = 0x0002C000
  SDRAM0_DATA_MEM (RW)   : origin = 0x02000000 length = 0x00400000
  SDRAM1_DATA_MEM (RW)   : origin = 0x03000000 length = 0x00400000
  INT_DATA_MEM    (RW)   : origin = 0x80000000 length = 0x00010000
}
```

132 C6x-Based DSP

```
          SECTIONS
          {
           .vec:         load = 0x00000000
           .text:        load = SBSRAM_PROG_MEM
           .const:       load = INT_DATA_MEM
           .bss:         load = INT_DATA_MEM
           .data:        load = INT_DATA_MEM
           .cinit        load = INT_DATA_MEM
           .pinit        load = INT_DATA_MEM
           .stack        load = INT_DATA_MEM
           .far          load = INT_DATA_MEM
           .sysmem       load = SDRAM0_DATA_MEM
           .cio          load = INT_DATA_MEM
           .fft          load = INT_PROG_MEM
           sbsbuf        load = SBSRAM_DATA_MEM
                   { _SbsramDataAddr = .; _SbsramDataSize = 0x0002C000; }
          }
```

Main program for FFT: *evm_fft.c*

```c
#include <stdlib.h>
#include <stdio.h>
#include <common.h>
#include <mcbspdrv.h>
#include <intr.h>
#include <board.h>
#include <codec.h>
#include <mcbsp.h>
#include <dma.h>
#include <math.h>
#include <regs.h>

#include "twiddleR.h"
#include "twiddleI.h"

#pragma DATA_ALIGN(x,64);

typedef struct
{
        short imag;
        short real;
} COEFF;

typedef struct
{
        short real;
        short imag;
} COMPLEX;
```

Appendix B   FFT Program Listing for Lab 6   **133**

```c
#define   NUMDATA      128                    // number of real data samples
#define   NUMPOINTS    NUMDATA/2              // number of point in the DFT
#define   PI           3.141592653589793
#define   BUFFLENGTH   128                    //for data acquisition

COEFF W4[64];
COMPLEX x[NUMPOINTS+1]; /* array of complex DFT data */

COMPLEX A[NUMPOINTS]; /* array of complex A coefficients */
COMPLEX B[NUMPOINTS]; /* array of complex B coefficients */
COMPLEX G[2*NUMPOINTS]; /* array of complex DFT result  */
unsigned short IIndex[NUMPOINTS], JIndex[NUMPOINTS];
int count;
short g[128];
int mag1[NUMDATA];
int mag2[NUMDATA];
int mag[NUMDATA];

short *output;
short *input;
short *intermediate;    /* POINTER TO DATA ARRAY FOR DMA ACCESS    */

int index=0,S1=20,S2=20;

void R4DigitRevIndexTableGen(int, int *, unsigned short *, unsigned short *);
void split1(int, COMPLEX *, COMPLEX *, COMPLEX *, COMPLEX *);
void digit_reverse(int *, unsigned short *, unsigned short *, int);
void radix4(int, short[], short[]);
void make_q15(short out[], float in[], int N);
void rescale(short g[]);
void hookint(void);
interrupt void serialPortRcvISR (void);
void WriteFIFO_DMA(void);
void wait_buffer(void);
void init_arrays(void);
void init_serial(void);

Mcbsp_dev dev;
Mcbsp_config   mcbspConfig;
int sampleRate,status,actualrate;


main()
{
        int i,k,n;
        short tr[64], ti[64];

        evm_init();
        make_q15(tr, TR, 64);   //Data in Header files from Matlab
        make_q15(ti, TI, 64);

        for(n=0;n<64;n++)
```

```c
        {
                W4[n].real = tr[n];
                W4[n].imag = ti[n];
        }

        for(k=0; k<NUMPOINTS; k++)
        {
          A[k].imag = (short)(16383.0*(-cos(2*PI/(double)(2*NUMPOINTS)*(double)k)));
          A[k].real = (short)(16383.0*(1.0 - sin(2*PI/(double)(2*NUMPOINTS)*(double)k)));
          B[k].imag = (short)(16383.0*(cos(2*PI/(double)(2*NUMPOINTS)*(double)k)));
          B[k].real = (short)(16383.0*(1.0 + sin(2*PI/(double)(2*NUMPOINTS)*(double)k)));
        }

        R4DigitRevIndexTableGen(NUMPOINTS, &count, IIndex, JIndex);

        mcbsp_drv_init();
        dev = mcbsp_open(0);
        init_serial();
        init_arrays();

        for(;;)
        {
                wait_buffer();
        }
}


/***************/
/*  FUNCTIONS  */
/***************/
void wait_buffer(void)
{
        int n,k;
        short *p;

        while(index);
        p = input;
        input = output;
        output = intermediate;

        for (n=0; n<NUMPOINTS; n++)
        {
                x[n].imag = p[2*n + 1];    // x2(n) = g(2n + 1)
                x[n].real = p[2*n];        // x1(n) = g(2n)
        }
```

```
            radix4(NUMPOINTS, (short *)x, (short *)W4);
            digit_reverse((int *)x, IIndex, JIndex, count);
            x[NUMPOINTS].real = x[0].real;
            x[NUMPOINTS].imag = x[0].imag;
            split1(NUMPOINTS, x, A, B, G);
            G[NUMPOINTS].real = x[0].real - x[0].imag;
            G[NUMPOINTS].imag = 0;
            for (k=1; k<NUMPOINTS; k++)
            {
                    G[2*NUMPOINTS-k].real = G[k].real;
                    G[2*NUMPOINTS-k].imag = -G[k].imag;
            }
            for (k=1; k<NUMDATA; k++)
            {
                    mag1[k] = (G[k].real*G[k].real) << 1;
                    mag2[k] = (G[k].imag*G[k].imag) << 1;
                    mag[k] = mag1[k] + mag2[k];
            }
            intermediate = p;
            WriteFIFO_DMA();
            while(!index);
}

void init_arrays(void)
{
  input        = (short *) calloc(BUFFLENGTH, sizeof(short)); /* AIC INPUT   */
  output       = (short *) calloc(BUFFLENGTH, sizeof(short)); /* AIC OUTPUT  */
  intermediate = (short *) calloc(BUFFLENGTH, sizeof(short)); /* DMA ACCESS  */
}




void hookint()
{
        intr_init();
        intr_map(CPU_INT15, ISN_RINT0);
        intr_hook(serialPortRcvISR, CPU_INT15);
        INTR_ENABLE(15);
        INTR_GLOBAL_ENABLE();
        return;
}
```

```c
void init_serial(void)
{
        memset(&mcbspConfig,0,sizeof(mcbspConfig));
        mcbspConfig.loopback            = FALSE;
        mcbspConfig.tx.update           = TRUE;
        mcbspConfig.tx.clock_mode       = CLK_MODE_EXT;
        mcbspConfig.tx.frame_length1    = 0;
        mcbspConfig.tx.word_length1     = WORD_LENGTH_32;

        mcbspConfig.rx.update           = TRUE;
        mcbspConfig.rx.clock_mode       = CLK_MODE_EXT;
        mcbspConfig.rx.frame_length1    = 0;
        mcbspConfig.rx.word_length1     = WORD_LENGTH_32;
        mcbsp_config(dev,&mcbspConfig);
        MCBSP_ENABLE(0, MCBSP_BOTH);
        codec_init();
        codec_adc_control(LEFT,0.0,FALSE,LINE_SEL);
        codec_adc_control(RIGHT,0.0,FALSE,LINE_SEL);

        /* mute (L/R)LINE input to mixer              */
        codec_line_in_control(LEFT,MIN_AUX_LINE_GAIN,TRUE);
        codec_line_in_control(RIGHT,MIN_AUX_LINE_GAIN,TRUE);

        /* D/A 0.0 dB atten, do not mute DAC outputs   */
        codec_dac_control(LEFT, 0.0, FALSE);
        codec_dac_control(RIGHT, 0.0, FALSE);

        sampleRate = 8000;

        actualrate = codec_change_sample_rate(sampleRate, TRUE);

        codec_interrupt_enable();

        hookint();
}


void WriteFIFO_DMA(void)
{
        dma_reset();
        dma_init(2,                             //Channel
            0x0A000110u,                        //Primary Control Register (Peripherals pp4-9)
            0x0000000Au,                        //Secondary Control Register
            (unsigned int) intermediate,        //Source Address
            0x01710000u,                        //Destination Address
             0x00010080u);                      //Transfer Counter Register

        DMA_START(DMA_CH2);
}
```

Appendix B FFT Program Listing for Lab 6

```c
interrupt void serialPortRcvISR (void)
{
        int temp;
        temp = MCBSP_READ(0);
        input[index] = temp >> S1;
        MCBSP_WRITE(0, output[index] << S2);
        if (++index == BUFFLENGTH)
                index = 0;
}

void make_q15(short out[], float in[], int N)
{
        int i;

        for(i=0;i<N;i++)
        {
                out[i]=0x7fff*in[i];   //Convert to Q-15, good approximate
        }
}

void radix4(int n, short x[], short w[])
{
        nt n1, n2, ie, ia1, ia2, ia3, i0, i1, i2, i3, j, k;
        short t, r1, r2, s1, s2, co1, co2, co3, si1, si2, si3;
        n2 = n;
        ie = 1;
        for (k = n; k > 1; k >>= 2) {
                n1 = n2;
                n2 >>= 2;
                ia1 = 0;
                for (j = 0; j < n2; j++) {
                        ia2 = ia1 + ia1;
                        ia3 = ia2 + ia1;
                        co1 = w[ia1 * 2 + 1];
                        si1 = w[ia1 * 2];
                        co2 = w[ia2 * 2 + 1];
                        si2 = w[ia2 * 2];
                        co3 = w[ia3 * 2 + 1];
                        si3 = w[ia3 * 2];
                        ia1 = ia1 + ie;
                        for (i0 = j; i0 < n; i0 += n1) {
                                i1 = i0 + n2;
                                i2 = i1 + n2;
                                i3 = i2 + n2;
                                r1 = x[2 * i0] + x[2 * i2];
                                r2 = x[2 * i0] - x[2 * i2];
                                t = x[2 * i1] + x[2 * i3];
                                x[2 * i0] = r1 + t;
                                r1 = r1 - t;
```

```
                    s1 = x[2 * i0 + 1] + x[2 * i2 + 1];
                    s2 = x[2 * i0 + 1] - x[2 * i2 + 1];
                    t  = x[2 * i1 + 1] + x[2 * i3 + 1];
                    x[2 * i0 + 1] = s1 + t;
                    s1 = s1 - t;
                    x[2 * i2] = (r1 * co2 + s1 * si2) >> 15;
                    x[2 * i2 + 1] = (s1 * co2-r1 * si2)>>15;
                    t = x[2 * i1 + 1] - x[2 * i3 + 1];
                    r1 = r2 + t;
                    r2 = r2 - t;
                    t = x[2 * i1] - x[2 * i3];
                    s1 = s2 - t;
                    s2 = s2 + t;
                    x[2 * i1] = (r1 * co1 + s1 * si1) >>15;
                    x[2 * i1 + 1] = (s1 * co1-r1 * si1)>>15;
                    x[2 * i3] = (r2 * co3 + s2 * si3) >>15;
                    x[2 * i3 + 1] = (s2 * co3-r2 * si3)>>15;
                }
            }
            ie <<= 2;
        }
}

void digit_reverse(int *yx, unsigned short *JIndex, unsigned short *IIndex, int count)
{
        int i;
        unsigned short I, J;
        int YXI, YXJ;
        for (i = 0; i<count; i++)
        {
                I = IIndex[i];
                J = JIndex[i];
                YXI = yx[I];
                YXJ = yx[J];
                yx[J] = YXI;
                yx[I] = YXJ;
        }
}




void R4DigitRevIndexTableGen(int n, int *count, unsigned short *IIndex, unsigned short *JIndex)
{
        int j, n1, k, i;
        j = 1;
        n1 = n - 1;
        *count = 0;
```

```c
            for(i=1; i<=n1; i++)
            {
                    if(i < j)
                    {
                            IIndex[*count] = (unsigned short)(i-1);
                            JIndex[*count] = (unsigned short)(j-1);
                            *count = *count + 1;
                    }

                    k = n >> 2;

                    while(k*3 < j)
                    {
                            j = j - k*3;
                            k = k >> 2;
                    }

                    j = j + k;
            }
}
void split1(int N, COMPLEX *X, COMPLEX *A, COMPLEX *B, COMPLEX *G)
{
          int k;
          int Tr, Ti;

          for (k=0; k<N; k++)
          {
                    Tr = (int)X[k].real * (int)A[k].real - (int)X[k].imag * (int)A[k].imag +
                            (int)X[N-k].real * (int)B[k].real + (int)X[N-k].imag * (int)B[k].imag;
                    G[k].real = (short)(Tr>>15);

                    Ti = (int)X[k].imag * (int)A[k].real + (int)X[k].real * (int)A[k].imag +
                            (int)X[N-k].real * (int)B[k].imag - (int)X[N-k].imag * (int)B[k].real;
                    G[k].imag = (short)(Ti>>15);
          }
}
```

APPENDIX C

Adaptive Filter Program Listing for Lab 7

Main C sampling: *adapt.c*

```c
#include <stdlib.h>
#include <string.h>
#include <common.h>
#include <mcbspdrv.h>
#include <intr.h>
#include <board.h>
#include <codec.h>
#include <mcbsp.h>

//IIR filter coefficients
float a[7] = {0.0179,0.9409,0.0104,0.6601,0.0342,0.1129,0.0058};
float b[8] = {0.1191,0.0123,-0.1813,-0.0251,0.1815,0.0307,-0.1194,-0.0178};

short AS[7];
short BS[8];

int S = 16;

short IIRwindow[8] = {0,0,0,0,0,0,0,0};
short FIRwindow[32] =
{0,0,0,0,0,0,0,0,0,0,0,0,0,0,0,0,0,0,0,0,0,0,0,0,0,0,0,0,0,0,0,0};

short y_prev[7] = {0,0,0,0,0,0,0};

short y=0,
                    e;

short h[32]={0,0,0,0,0,0,0,0,0,0,0,0,0,0,0,0,0,0,0,0,0,0,0,0,0,0,0,0,0,0,0,0};

short DELTA = 0x0001;

void hookint(void);
extern interrupt void serialPortRcvISR (void);
```

Appendix C   Adaptive Filter Program Listing for Lab 7

```c
main()
{
        Mcbsp_dev dev;
        Mcbsp_config  mcbspConfig;
        int i,j,sampleRate,status,actualrate;


        for(i=0;i<7;i++)
        {
                AS[i]=0x7FFF * a[i];
                BS[i]=0x7FFF * b[i];
        }

        BS[7]=0x7FFF * b[7];


        /*******************************************/
        /* Initialize EVM                          */
        /*******************************************/
        evm_init();
        mcbsp_drv_init();
        dev = mcbsp_open(0);

        /*************************************************/
        /* configure McBSP                               */
        /*************************************************/

        memset(&mcbspConfig,0,sizeof(mcbspConfig));
        mcbspConfig.loopback            = FALSE;
        mcbspConfig.tx.update           = TRUE;
        mcbspConfig.tx.clock_mode       = CLK_MODE_EXT;
        mcbspConfig.tx.frame_length1    = 0;
        mcbspConfig.tx.word_length1     = WORD_LENGTH_32;
        mcbspConfig.rx.update           = TRUE;
        mcbspConfig.rx.clock_mode       = CLK_MODE_EXT;
        mcbspConfig.rx.frame_length1    = 0;
        mcbspConfig.rx.word_length1     = WORD_LENGTH_32;
        mcbsp_config(dev,&mcbspConfig);

        MCBSP_ENABLE(0, MCBSP_BOTH);

        /*************************************************/
        /* configure CODEC                               */
        /*************************************************/
        codec_init();

        /* A/D 0.0 dB gain, turn on 20dB mic gain, sel (L/R)LINE input     */
        codec_adc_control(LEFT,0.0,FALSE,LINE_SEL);
        codec_adc_control(RIGHT,0.0,FALSE,LINE_SEL);
```

142   C6x-Based DSP

```c
        /* mute (L/R)LINE input to mixer           */
        codec_line_in_control(LEFT,MIN_AUX_LINE_GAIN,TRUE);
        codec_line_in_control(RIGHT,MIN_AUX_LINE_GAIN,TRUE);

        /* D/A 0.0 dB atten, do not mute DAC outputs    */
        codec_dac_control(LEFT, 0.0, FALSE);
        codec_dac_control(RIGHT, 0.0, FALSE);

        sampleRate = 8000;

        actualrate = codec_change_sample_rate(sampleRate, TRUE);

        codec_interrupt_enable();

        hookint();

        /******************************************************************/
        /* Main Loop: wait for Interrupt                                  */
        /******************************************************************/
        while (1)
        {
        }
}
/*----------------------------------------------------------*/
/* FUNCTIONS                                                */
/*----------------------------------------------------------*/
void hookint()
{
        intr_init();
        intr_map(CPU_INT15, ISN_RINT0);
        intr_hook(serialPortRcvISR, CPU_INT15);

        INTR_ENABLE(15);
        INTR_GLOBAL_ENABLE();

        return;
}
```

IIR and FIR interrupt in C: *intr.c*

```c
#include <mcbspdrv.h>

extern short AS[7];
extern short BS[8];
extern int S;
extern short IIRwindow[8];
extern short FIRwindow[32];
extern short y_prev[7];
extern short y,e;
extern short h[32];
extern short DELTA;
```

Appendix C Adaptive Filter Program Listing for Lab 7

```c
interrupt void serialPortRcvISR (void)
{
        int mu,zemp,temp,temp2,n,ASUM,BSUM;
        short input,IIR_OUT;
        short output,stemp,stemp2;

        temp = MCBSP_READ(0);

        input = temp >> S;

        /*******************************/
        for(n=7;n>0;n--)
                IIRwindow[n] = IIRwindow[n-1];

        IIRwindow[0] = input;

        BSUM = 0;
        for(n=0;n<7;n++)
        {
                BSUM += ((BS[n] * IIRwindow[n]) << 1);
        }

        ASUM = 0;
        for(n=0;n<6;n++)
        {
                ASUM += ((AS[n] * y_prev[n]) << 1);
        }

        IIR_OUT = (BSUM - ASUM) >> 16;

        for(n=6;n>0;n--)
                y_prev[n] = y_prev[n-1];

        y_prev[0] = IIR_OUT;

        /*************************/
        /*ADAPTIVE FIR FILTERING*/
        /*************************/
        /*Seperate Circular buffer for FIR*/
        for(n=31;n>0;n--)
                FIRwindow[n] = FIRwindow[n-1];

        FIRwindow[0] = input;

        //Perform Filtering with current coefficients
        zemp = 0;
        for(n=0;n<32;n++)
        {
                zemp += ((h[n]*FIRwindow[n]) << 1);
        }
```

```c
y = zemp >> 16;

//Calculate Error Term

e = IIR_OUT - y;

//Update Coefficients

stemp = (DELTA*e)>>15;

for(n=0;n<32;n++)
{

        stemp2 = (stemp*FIRwindow[n])>>15;

        h[n] = h[n] + stemp2;
}

MCBSP_WRITE(0, y << S);

}
```

Linker command file: *linkADAPT.cmd*

```
MEMORY
{
  INT_PROG_MEM   (RX)   : origin = 0x00000000 length = 0x00010000
  SBSRAM_PROG_MEM (RX)  : origin = 0x00400000 length = 0x00014000
  SBSRAM_DATA_MEM (RW)  : origin = 0x00414000 length = 0x0002C000
  SDRAM0_DATA_MEM (RW)  : origin = 0x02000000 length = 0x00400000
  SDRAM1_DATA_MEM (RW)  : origin = 0x03000000 length = 0x00400000
  INT_DATA_MEM   (RW)   : origin = 0x80000000 length = 0x00010000
}

SECTIONS
{
  .vec:       load = 0x00000000
  .text:      load = SBSRAM_PROG_MEM
  .const:     load = INT_DATA_MEM
  .bss:       load = INT_DATA_MEM
  .data:      load = INT_DATA_MEM
  .cinit      load = INT_DATA_MEM
  .pinit      load = INT_DATA_MEM
  .stack      load = INT_DATA_MEM
  .far        load = INT_DATA_MEM
  .sysmem     load = SDRAM0_DATA_MEM
  .cio        load = INT_DATA_MEM
  .int        load = INT_PROG_MEM
  sbsbuf      load = SBSRAM_DATA_MEM
           { _SbsramDataAddr = .; _SbsramDataSize = 0x0002C000; }
}
```

Appendix C Adaptive Filter Program Listing for Lab 7

Interrupt in C with IIR and call to FIR in assembly: *adapt.c* (C+ASM directory)

```
interrupt void serialPortRcvISR (void)
{
    int temp,n,ASUM,BSUM;
    short input,IIR_OUT,output;

        temp = MCBSP_READ(0);

        input = temp >> S;

        /*****************************/
    for(n=7;n>0;n--)
        IIRwindow[n] = IIRwindow[n-1];


    IIRwindow[0] = input;


    BSUM = 0;
    for(n=0;n<7;n++)
    {
        BSUM += ((BS[n] * IIRwindow[n]) << 1);
    }


    ASUM = 0;
    for(n=0;n<6;n++)
    {
        ASUM += ((AS[n] * y_prev[n]) << 1);
    }


    IIR_OUT = (BSUM - ASUM) >> 16;


    for(n=6;n>0;n--)
        y_prev[n] = y_prev[n-1];


    y_prev[0] = IIR_OUT;


    output = adaptFIR(input,IIR_OUT);
    //output = IIR_OUT;

        MCBSP_WRITE(0, output << S);


}
```

146   C6x-Based DSP

Adaptive filter in assembly using circular buffering: *adapt.asm*

```
;A4=Passed sample Value, B4=Passed d[n] value (IIR_OUT)
        .global _adaptFIR

        .sect   ".text"

_adaptFIR:

        ;Initialize the Circular buffer for the FIR filter
        MVK     .S2     0x0004,B10      ;A5 is selected as circular
        MVKLH   .S2     0x0005,B10      ;2^(4+1)=32
        MVC     .S2     B10,AMR

        ;Load the pointer to A1
        ;Assume that the current location of the circular buffer
        ;is pointed to

        MVK     .S1     0x0080,A0
        MVKLH   .S1     0x8000,A0       ;A0=0x80000080 (has last pointer)

        LDW     .D1     *A0,A5
        NOP     4
                                        ;A5 now points to the first free
                                        ;Location of the Circular buffer
        ;Load the current sample to the Circular buffer
        STH     .D1     A4,*A5          ;A4 has sample passed from Calling

        ;Do the filtering
        MVK     .S2     0x0040,B1
        MVKLH   .S2     0x8000,B1       ;This is the address of h[n]

        MVK     .S2     32,B2   ;Set up a counter
        ZERO    .S1     A9                      ;Accumulator

loop:
        LDH     .D2     *B1++,B7        ;load hk
        LDH     .D1     *A5-,A7
        NOP     4
        MPY     .M1x    A7,B7,A7        ;A7 is Q-30
        NOP
        SHL     .S1     A7,1,A7
        ADD     .S1     A7,A9,A9
 [B2]   SUB     .S2     B2,1,B2         ;Decrement Counter
 [B2]   B       .S2     loop
        NOP     5

        SHR     .S1     A9,16,A9        ;Make Short,Eliminate Sign extension bit
                                        ;A9 is now short Y
```

```
;Calculate Error Term
        MV      .S1     B4,A13
        SUB     .S1     A13,A9,A8       ;A13=d(IIR_OUT),A9=y,A8=e

;Update Coefficients
        MVK     .S1     0x0001,A10      ;A10=DELTA
        MPY     .M1     A8,A10,A10      ;A10=DELTA*e this ia actually ineffective
        NOP
        SHR     .S1     A10,15,A10      ;A10=DELTA*e is now short Q-15

        MVK     .S2     32,B2           ;Loop Counter
        MVK     .S2     0x0040,B1
        MVKLH.S2        0x8000,B1       ;This is the address of h[n]

loop2:

        LDH     .D1     *A5-,A8         ;Load x[n-k]
        LDH     .D2     *B1,A12         ;Load h[n]
        NOP             4
        MPY     .M1     A10,A8,A8       ;A10 = DELTA*e*x in Q-31
        NOP
        SHR     .S1     A8,15,A8        ;A10 is now Q-15
        ADD     .S1     A8,A12,A8       ;Updated h
        STH     .D2     A8,*B1++        ;Update the coefficient
[B2]    SUB     .S2     B2,1,B2;Decrement Counter
[B2]    B       .S2     loop2
        NOP             5

;Now save the Last location of A1 to memory

        MVK     .S1     0x0080,A0
        MVKLH.S1        0x8000,A0       ;This address has the pointer to x

        LDH     .D1     *A5++,A13       ;Dummy Load
        STW     .D1     A5,*A0          ;Saved last




;Restore Linear Addressing
        MVK     .S2     0x0000,B10
        MVKLH.S2        0x0004,B10
        MVC     .S2     B10,AMR

;return the result y

        MV      .S1     A9,A4
        B       .S2     B3
        NOP             5
```

148 C6x-Based DSP

Command file for assembly version: *link.cmd* (C+ASM and C+SA directories)

```
MEMORY
{
 INT_PROG_MEM    (RX)  : origin = 0x00000000 length = 0x00010000
 SBSRAM_PROG_MEM (RX)  : origin = 0x00400000 length = 0x00014000
 SBSRAM_DATA_MEM (RW)  : origin = 0x00414000 length = 0x0002C000
 SDRAM0_DATA_MEM (RW)  : origin = 0x02000000 length = 0x00400000
 SDRAM1_DATA_MEM (RW)  : origin = 0x03000000 length = 0x00400000
 INT_DATA_MEM    (RW)  : origin = 0x80000100 length = 0x0000FF00
 MMEM                  : origin = 0x80000000 length = 0x00000100
}

SECTIONS
{
 .vec:       load = 0x00000000
 .text:      load = SBSRAM_PROG_MEM
 .const:     load = INT_DATA_MEM
 .bss:       load = INT_DATA_MEM
 .data:      load = INT_DATA_MEM
 .cinit      load = INT_DATA_MEM
 .pinit      load = INT_DATA_MEM
 .stack      load = INT_DATA_MEM
 .far        load = INT_DATA_MEM
 .sysmem     load = SDRAM0_DATA_MEM
 .cio        load = INT_DATA_MEM
 .int        load = INT_PROG_MEM
 .mydata     load = MMEM
 sbsbuf      load = SBSRAM_DATA_MEM
        { SbsramDataAddr = .; _SbsramDataSize = 0x0002C000; }
}
```

Linear assembly file: *adaptFIR.sa*

```
;A4=Passed sample Value, B4=Passed d[n] value (IIR_OUT)
        .global    _adaptFIR

        .sect      ".text"

_adaptFIR: .proc    a4,b4,b3

   .reg x,d,temp,temp1,temp2,temp3,ptr,ph,count,delta,sum,error,ED,XL,H;


       mv a4,x
       mv b4,d
```

Appendix C Adaptive Filter Program Listing for Lab 7

```
        ;Initialize the Circular buffer for the FIR filter
        MVK     0x0004,temp     ;A5 is selected as circular
        MVKLH   0x0005,temp     ;2^(4+1)=32
        MVC     temp,AMR

        ;Load the pointer to A1
        ;Assume that the current location of the circular buffer
        ;is pointed to

        MVK     0x0080,temp
        MVKLH   0x8000,temp     ;A0=0x80000080 (has last pointer)

        LDW     *temp,ptr
        NOP     4               ;A5 now points to the first free
                                ;Location of the Circular buffer

        ;Load the current sample to the Circular buffer
        STH     x,*ptr          ;x has sample passed from Calling

        ;Do the filtering
        MVK     0x0040,ph
        MVKLH   0x8000,ph       ;This is the address of h[n]

        MVK     32,count        ;Set up a counter
        ZERO    sum             ;Accumulator

loop:   .trip 32

        LDH     *ph++,temp1     ;load hk
        LDH     *ptr-,temp2
        NOP     4
        MPY     temp1,temp2,temp1  ;A7 is Q-30
        NOP
        SHL     temp1,1,temp1
        ADD     temp1,sum,sum
[count] SUB     count,1,count   ;Decrement Counter
[count] B       loop
        NOP     5

        SHR     sum,16,sum      ;Make Short,Eliminate Sign extension bit
                                ;A9 is now short Y

        ;Calculate Error Term

        SUB     d,sum,error     ;A13=d(IIR_OUT),A9=y,A8=e

        ;Update Coefficients
        MVK     0x0100,delta    ;A10=DELTA
        MPY     error,delta,ED  ;A10=DELTA*e
        NOP
        SHR     ED,15,ED        ;A10=DELTA*e is now short Q-15
```

150 C6x-Based DSP

```
            MVK     32,count            ;Loop Counter
            MVK     0x0040,ph
            MVKLH           0x8000,ph   ;This is the address of h[n]

loop2:      .trip 32

            LDH     *ptr-,XL            ;Load x[n-k]
            LDH     *ph,H               ;Load h[n]
            NOP     4
            MPY     ED,XL,XL            ;A10 = DELTA*e*x in Q-31
            NOP
            SHR     XL,15,XL            ;A10 is now Q-15
            ADD     XL,H,XL             ;Updated h
            STH     XL,*ph++            ;Update the coefficient
[count]     SUB     count,1,count       ;Decrement Counter
[count]     B       loop2
            NOP     5

;Now save the Last location of A1 to memory

            MVK     0x0080,temp
            MVKLH           0x8000,temp     ;This address has the
pointer to x

            LDH     *ptr++,temp3        ;Dummy Load
            STW     ptr,*temp           ;Saved last

            ;Restore Linear Addressing
            MVK     0x0000,temp2
            MVKLH           0x0004,temp2
            MVC     temp2,AMR

            ;return the result y
            MV      sum,a4

            .endproc a4,b3

            B       B3
            NOP     5
```

APPENDIX D

Quick Reference Guide

D–1: LIST OF C6X INSTRUCTIONS

.L Unit	
Instruction	Description
ABS	Integer absolute value with saturation
ADD(U)	Signed(unsigned) addition without saturation
AND	Bitwise AND
CMPEQ	Integer compare for equality
CMPGT	Signed integer compare for greater than
CMPGTU	Unsigned integer compare for greater than
CMPLT	Signed integer compare for less than
CMPLTU	Unsigned integer compare for less than
LMBD	Leftmost bit detection
MV	Move from register to register
NEG	Negate
NORM	Normalize integer
NOT	Bitwise NOT
OR	Bitwise OR
SADD	Integer addition with saturation to result size
SAT	Saturate a 40-bit integer to a 32-bit integer
SSUB	Integer subtraction with saturation to result size
SUBU	Unsigned integer subtraction without saturation
SUBC	Conditional integer subtract and shift (used for division)
XOR	Exclusive OR
ZERO	Zero a register

.M Unit	
MPY	Signed integer multiply 16 lsb $\times$ 16 lsb
MPYU	Unsigned integer multiply 16 lsb $\times$ 16 lsb
MPYUS	Integer multiply (unsigned) 16 lsb $\times$ (signed) 16 lsb
MPYSU	Integer multiply (signed) 16 lsb $\times$ (unsigned) 16 lsb
MPYH	Signed integer multiply 16 msb $\times$ 16 msb
MPYHU	Unsigned integer multiply 16 msb $\times$ 16 msb
MPYHUS	Integer multiply (unsigned) 16 msb $\times$ (signed) 16 msb
MPYHSU	Integer multiply (signed) 16 msb $\times$ (unsigned) 16 msb
MPYHL	Signed multiply high low 16 msb $\times$ 16 lsb
MPYHLU	Unsigned multiply high low 16 msb $\times$ 16 lsb

MPYHULS	Multiply high unsigned low signed (unsigned) 16 msb × (signed) 16 lsb
MPYHSLU	Multiply high signed low unsigned (signed) 16 msb × (unsigned) 16 lsb
SMPY	Integer multiply with left shift and saturation
SMPYHL	Integer multiply high low with left shift and saturation
SMPYLH	Integer multiply low high with left shift and saturation
SMPYH	Integer multiply high with left shift and saturation

.S Unit

ADD(U)	Signed(unsigned) addition without saturation
ADDK	Integer addition using signed 16-bit constant
ADD2	Two 16-bit integer adds on upper and lower register halves
AND	Bitwise AND
B disp	Branch using a displacement
B IRP	Branch using an Interrupt return pointer
B NRP	Branch using a NMI return pointer
B reg	Branch using a register
CLR	Clear a bit field
EXT(U)	Extract and sign-extend(zero-extend) a bit field
MV	Move from register to register
MVC	Move between the control file and register file
MVK	Move a 16-bit signed constant into a register and sign extend
MVKH	Move 16-bit constant into the upper bits of a register
MVKLH	Move 16-bit constant into the upper bits of a register
NEG	Negate
NOT	Bitwise NOT
OR	Bitwise OR
SET	Set a bit field
SHL	Arithmetic shift left
SHR	Arithmetic shift right
SHRU	Logical shift right
SHRL	Logical shift right
SUB(U)	Signed(Unsigned) integer subtraction without saturation
SUB2	Two 16-bit integer subtracts on upper and lower register halves
XOR	Exclusive OR
ZERO	Zero a register

.D Unit

ADD(U)	Signed(unsigned) integer addition without saturation
ADDAB/ADDAH/ADDAW	Integer addition using addressing mode
LDB(U)/LDH(U)/LDW	Load from memory with a 5-bit unsigned constant offset or register offset
LDB(U)/LDH(U)/LDW (15-bit offset)	Load from memory with a 15-bit constant offset
MV	Move from register to register
STB/STH/STW	Store to memory with a register offset or 5-bit unsigned constant offset
STB/STH/STW (15-bit offset)	Store to memory with a 15-bit offset
SUB	Signed integer subtraction without saturation
SUBAB/SUBAH/SUBAW	Integer subtraction using addressing mode
ZERO	Zero a register

D–2 REGISTERS AND MEMORY-MAPPED REGISTERS†

Addressing Mode Register (AMR)

31	26	25	21	20	16
Reserved		BK1		BK0	
R,+0		R,W,+0		R,W,+0	

15	14	13	12	11	10	9	8	7	6	5	4	3	2	1	0
B7 mode		B6 mode		B5 mode		B4 mode		A7 mode		A6 mode		A5 mode		A4 mode	

R,W,+0

Control Status Register (CSR)

31	24	23	16
CPUID		Revision ID	
R		R,W,+0	

15	10	9	8	7	5	4	2	1	0
PWRD		SAT	EN	PCC		DCC		PGIE	GIE
R,W,+0		R,C+0	R,+x			R,W,+0			

Interrupt Flag Register (IFR)

15	14	13	12	11	10	9	8	7	6	5	4	3	2	1	0
IF15	IF14	IF13	IF12	IF11	IF10	IF9	IF8	IF7	IF6	IF5	IF4	rsv	rsv	NMIF	0

R,+0

Interrupt Set Register (ISR)

15	14	13	12	11	10	9	8	7	6	5	4	3	2	1	0
IS15	IS14	IS13	IS12	IS11	IS10	IS9	IS8	IS7	IS6	IS5	IS4	rsv	rsv	rsv	rsv

W

Interrupt Clear Register (ICR)

15	14	13	12	11	10	9	8	7	6	5	4	3	2	1	0
IC15	IC14	IC13	IC12	IC11	IC10	IC9	IC8	IC7	IC6	IC5	IC4	rsv	rsv	rsv	rsv

W

Interrupt Enable Register (IER)

15	14	13	12	11	10	9	8	7	6	5	4	3	2	1	0
IE15	IE14	IE13	IE12	IE11	IE10	IE9	IE8	IE7	IE6	IE5	IE4	rsv	rsv	NMIE	1
R,W,+0															R,1

Interrupt Service Table Pointer (ISTP)

31	10	9	5	4	3	2	1	0
ISTB		HPEINT		0	0	0	0	0
R,W,+0		R,+0						

NMI Return Pointer (NRP)

31	0
NRP	
R,W,+x	

Interrupt Return Pointer (IRP)

31	0
IRP	
R,W,+x	

note: bits not shown are **reserved**

EMIF

CE0 Space Control	1800008
CE1 Space Control	1800004
CE2 Space Control	1800010
CE3 Space Control	1800014
Global Control	1800000
SDRAM Control	1800018
SDRAM Refresh Period	180001C

HPI

Control Register	1880000

Interrupts

Multiplexer High	19C0000
Multiplexer Low	19C0004
External Interrupt Polarity	19C0008

DMA

	Ch. 0	Ch. 1	Ch. 2	Ch. 3
Primary Control	1840000	1840040	1840004	1840044
Secondary Control	1840008	1840048	184000C	184004C
Source Address	1840010	1840050	1840014	1840054
Destination Address	1840018	1840058	184001C	184005C
Transfer Counter	1840020	1840060	1840024	1840064
Global Reload A	1840028			
Global Reload B	184002C			
Global Index A	1840030			
Global Index B	184003C			
Global Index C	1840068			
Global Index D	184006C			
Auxiliary Control	1840070			

McBSP

	0	1
DRR	18C0000	1900000
DXR	18C0004	1900004
Control Register	18C0008	1900008
Receive Control Register	18C000C	190000C
Transmit Control Register	18C0010	1900010
Sample Rate Generator Register	18C0014	1900014
Multichannel Register	18C0018	1900018
Receive Channel Enable Register	18C001C	190001C
Transmit Channel Enable Register	18C0020	1900020
Pin Control Register	18C0024	1900024

Timers

	0	1
Control	1940000	1980000
Period	1940004	1980004
Counter	1940008	1980008

C6x_MMR†

```
************************************************************************
*     Memory-Mapped Registers
*
*     You must include this file into each .asm that references
*     one of these mmr register names.  It can be included by adding
*     this line to the top of your .asm file:
*
*        .include c6x_mmr.asm
*
*     Using the names below simplifies access to peripheral mmr registers.
*     Here's an example for writing all F's into the CE1 and CE2 EMIF
*     space control registers:
*        MVK       .S1          0FFFFh, A0
*        MVKLH     .S1          0FFFFh, A0
*        MVK       .S1          EMIF, A1
*        MVKH      .S1          EMIF, A1
*        STW       .D1          A0, *+A1[CE1]
*        STW       .D1          A0, *+A1[CE2]
************************************************************************

;Peripheral      Addr/Offset           Register
;--------        ----------            ---------------------------------

EMIF            .equ   01800000h     ; EMIF global control
CE1             .equ   1             ; EMIF CE1 space control
CE0             .equ   2             ; EMIF CE0 space control
CE2             .equ   4             ; EMIF CE2 space control
CE3             .equ   5             ; EMIF CE3 space control
SDRAM           .equ   6             ; EMIF SDRAM control
REFRESH         .equ   7             ; EMIF SDRAM refresh period

DMA             .equ   01840000h     ; Top of DMA registers
DMA0pc          .equ   0             ; DMA0 primary control
DMA2pc          .equ   1             ; DMA2 primary control
DMA0sc          .equ   2             ; DMA0 secondary control
DMA2sc          .equ   3             ; DMA2 secondary control
DMA0src         .equ   4             ; DMA0 source address
DMA2src         .equ   5             ; DMA2 source address
DMA0dst         .equ   6             ; DMA0 destination address
DMA2dst         .equ   7             ; DMA2 destination address
DMA0tc          .equ   8             ; DMA0 transfer counter
DMA2tc          .equ   9             ; DMA2 transfer counter
DMAcountA       .equ   10            ; DMA global count reload register A
DMAcountB       .equ   11            ; DMA global count reload register B
```

```
DMAindexA    .equ   12           ; DMA global index register A
DMAindexB    .equ   13           ; DMA global index register B
DMAaddrA     .equ   14           ; DMA global address register A
DMAaddrB     .equ   15           ; DMA global address register B
DMA1pc       .equ   16           ; DMA1 primary control
DMA3pc       .equ   17           ; DMA3 primary control
DMA1sc       .equ   18           ; DMA1 secondary control
DMA3sc       .equ   19           ; DMA3 secondary control
DMA1src      .equ   20           ; DMA1 source address
DMA3src      .equ   21           ; DMA3 source address
DMA1dst      .equ   22           ; DMA1 destiantion address
DMA3dst      .equ   23           ; DMA3 destination address
DMA1tc       .equ   24           ; DMA1 transfer counter
DMA3tc       .equ   25           ; DMA3 transfer counter
DMAaddrC     .equ   26           ; DMA global address register C
DMAaddrD     .equ   27           ; DMA global address register D
DMAaux       .equ   28           ; DMA auxiliary control register

HPI          .equ   01880000h    ; HPI control register


McBSP0       .equ   018C0000h    ; McBSP0 DRR
McBSP1       .equ   01900000h    ; McBSP1 DRR
DRR          .equ   0            ; McBSP DRR
DXR          .equ   1            ; McBSP DXR
SPCR         .equ   2            ; McBSP control register
RCR          .equ   3            ; McBSP receive control register
XCR          .equ   4            ; McBSP transmit control register
SRGR         .equ   5            ; McBSP sample rate generator register
MCR          .equ   6            ; McBSP multicahnnel register
RCER         .equ   7            ; McBSP receive channel enable register
CER          .equ   8            ; McBSP transmit channel enable register
PCR          .equ   9            ; McBSP pin control register


Timer0       .equ   01940000h    ; Timer 0
Timer 1      .equ   01980000h    ; Timer 1
TimCR        .equ   0            : Timer control register
TimTP        .equ   1            ; Timer period
TimTC        .equ   2            ; Timer counter


Interrupts   .equ   019C0000h    ; Interrupts
IMH          .equ   0            ; Interrupt Multiplexer high
IML          .equ   1            ; Interrupt Multiplexer low
IP           .equ   2            ; External interrupt polarity
```

Appendix D   Quick Reference Guide   **157**

D-3: SIMULATOR QUICK REFERENCE†

Entry/Exit

- sim6x <file> Start simulator and loads <file>. Executed from the DOS prompt
- QUIT   Quit simulator

Working With Files

Loading
- LOAD <file>   Load <file>.out
- RELOAD <file>   Load object code of <file>
- SLOAD <file>   Load sybol table of <file>

Displaying
- FILE <file>   Display a text file

Entry/Exit

Reset
- RESET   Force PC to zero
- RESTART   Return to entry point

Run and Stop
- RUN or F5   Run a program
- GO <value>   Run to an address or symbol
- RUNB   Set CLK to zero and run
- RETURN   Return to a function's caller
- HALT or ESC   Halt target

Stepping
- STEP or F8   Single step
- CSTEP   Single step by C statement
- NEXT   Step over function calls
- CNEXT   Cstep over function calls

Watches and Breakpoints

Operation	Watch	Breakpoint
• ADD	WA<label>	BA<sym/addr>
• DELETE	WD<label>	BD #
• RESET	WR	BR
• LIST	WL	BL

Windows Management

Selecting Windows
- F6 — Select next window
- WIN <name> — Select <name> window
- Ctrl+F4 — Close selected window

Resizing Windows
- SIZE <name> — Resize <name> window

Moving Inside a Window
- Up Arrow/Down Arrow
- Page Up/Page Down
- For DISASSEM window, use ADDR <value>
- For MEMORY window, use MEM <value>

Screen Configuration
- SCONFIG <name> — Load config.<name>
- SSAVE <name> — Save config.<name>
- PROMPT <value> — Make prompt <value>

Modes
- ASM or ALT+D,A — Display ASM info
- C or ALT+D,C — Display C info
- MIX or ALT+D,M — Display BOTH

Other Actions

- ALT+1 — Go to simulator command line
- ?<label> — Display value of label
- ?<label>=<n> — Load <label> with <n>
- DLOG <file> — Record commands in a file
- DLOG CLOSE — Stop recording
- TAKE <file> — Execute batch file
- ALIAS — Record a new command string
- UNALIAS — Delete an alias definition
- FILL — Initialize a memory range
- Up/Down Key — Access previous commands

D-4: COMPILER INTRINSICS[†]

C Compiler Intrinsic	Assembly Instruction	Description	Device
int **_abs**(int *src2*);	**ABS**	Returns the saturated absolute value of *src2*	
int **_labs**(long *src2*);			
int **_add2**(int *src1*, int *src2*);	**ADD2**	Adds the upper and lower halves of *src1* to the upper and lower halves of *src2* and returns the result.	
uint **_clr**(uint *src2*, uint *csta*, uint *cstb*);	**CLR**	Clears the specified field in *src2*. The beginning and ending bits of the field to be cleared are specified by *csta* and *cstb*, respectively.	
unsigned **_clrr**(uint *src1*, int *src2*);	**CLR**	Clears the specified field in *src2*. The beginning and ending bits of the field to be cleared are specified by the lower 10 bits of the source register.	
int **_dpint**(double);	**DPINT**	Converts 64-bit double to 32-bit signed integer, using the rounding mode set by the CSR register	'C67x
int **_ext**(uint *src2*, uint *csta*, int *cstb*);	**EXT**	Extracts the specified field in *src2*, sign-extended to 32 bits. The extract is performed by a signed shift right; *csta* and *cstb* are the shift left and shift right amounts, respectively.	
int **_extr**(int *src2*, int *src1*);	**EXT**	Exctracts the specified field in *src2*, sign-extended to 32 bits.	
uint **_extu**(uint *src2*, uint *csta*, uint *cstb*);	**EXTU**	Extracts the specified field in *src2*, zero-extended to 32 bits.	
uint **_extur**(uint *src2*, int *src1*);	**EXTU**	Extracts the specified field in *src2*, zero-extended to 32 bits.	
uint **_ftoi**(float);		Reinterprets the bits in the float as an unsigned integer.	'C67x
uint **_hi**(double);		Returns the high 32 bits of a double as an integer.	'C67x
double **_itod**(uint, uint);		Creates a new double register pair from two unsigned integers	'C67x
float **_itof**(uint);		Reinterprets the bits in the unsigned integer as a float.	'C67x
uint **_lmbd**(uint *src1*, uint *src2*);	**LMBD**	Searches for a leftmost 1 or 0 of *src2* determined by the lsb of *src1*. Returns the number of bits up to the bit change.	
uint **_lo**(double);		Returns the low (even) register of a double register pair as an integer.	'C67x
int **_mpy**(int *src1*, int *src2*);	**MPY**	Multiplies the 16 lsbs of *src1* by the 16 MSBs of *src2* and returns the result. Values can be signed or unsigned.	
int **_mpyus**(uint *src1*, int *src2*);	**MPYUS**		
int **_mpysu**(int *src1*, uint *src2*);	**MPYSU**		
uint **_mpyu**(uint *src1*, uint *src2*);	**MPYU**		
int **_mpyhl**(int *src1*, int *src2*);	**MPYHL**	Multiplies the 16 msbs of *src1* by the 16 LSBs of *src2* and returns the result. Values can be signed or unsigned.	
int **_mpyhuls**(uint *src1*, int *src2*);	**MPYHULS**		
int **_mpyhslu**(int *src1*, uint *src2*);	**MPYHSLU**		
uint **_mpyhlu**(uint *src1*, uint *src2*);	**MPYHLU**		
int **_mpylh**(int *src1*, int *src2*);	**MPYLH**	Multiplies the 16 lsbs of *src1* by the 16	

int **_mpyluhs**(uint *src1*, int *src2*); int **_mpylshu**(int *src1*, uint *src2*); int **_mpylhu**(uint *src1*, uint *src2*); void **_nassert**(int);	**MPYLUHS** **MPYLSHU** **MPYLHU**	msbs of *src2* and returns the result. Values can be signed or unsigned. Generates no code. Tells the optimizer that the expression declared with the assert function is true; this gives a hint to the optimizer as to what optimizations might be valid.
uint **_norm**(int *src2*); uint **_lnorm**(long *src2*);	**NORM**	Returns the number of bits up to the first nonredundant sign bit of *src2*.
double **_rcpdp**(double);	**RCPDP**	Computes the approximate 64-bit double reciprocal. 'C67x
float **_rcpsp**(float);	**RCPSP**	Computes the approximate 32-bit float reciprocal. 'C67x
double **_rsqrdp**(float);	**RSQRDP**	Computes the approximate 64-bit double reciprocal square root. 'C67x
float **_rsqrsp**(float);	**RSQRSP**	Computes the approximate 32-bit float reciprocal square root. 'C67x
int **_sadd**(int *src1*, int *src2*); long **_lsadd**(int *src1*, long *src2*);	**SADD**	Adds *src1* to *src2* and saturates the results.
int **_sat**(long *src2*);	**SAT**	Converts a 40-bit value to an 32-bit value and saturates if necessary.
uint **_set**(uint *src2*, uint *csta*, uint *cstb*);	**SET**	Sets the specified field in *src2* to all 1s and returns the *src2* value. The beginning and ending bits of the field to be set are specified by *csta* and *cstb* respectively.
unsigned **_setr**(unsigned, int);	**SET**	Sets the specified field in *src2* to all 1s and returns the *src2* value. The beginning and ending bits of the field to be set are specified by the lower 10 bits of the source register.
int **_smpy**(int *src1*, int *src2*); int **_smpyh**(int *src1*, int *src2*); int **_smpyhl**(int *src1*, int *src2*); int **_smpylh**(int *src1*, int *src2*);	**SMPY** **SMPYH** **SMPYHL** **SMPYLH**	Multiplies *src1* by *src2*, left shifts the result by one, and returns the result. If the result is 0x8000 0000, saturates the result to 0x7FFF FFFF.
int **_spint**(float);	**SPINT**	Converts 32-bit float to 32-bit signed integer, using the rounding mode set by the CSR register. 'C67x
uint **_sshl**(uint *src2*, uint *src1*);	**SSHL**	Shifts *src2* left by the contents of *src1*, saturates the result to 32-bits, and returns the result.
int **_ssub**(int *src1*, uint *src2*); long **_lssub**(int *src1*, long *src2*);	**SSUB**	Subtracts *src2* from *src1*, saturates the result size, and returns the result.
uint **_subc**(uint *src1*, uint *src2*);	**SUBC**	Conditional subtract divide step.
int **_sub2**(int *src1*, int *src2*);	**SUB2**	Subtracts the upper and lower halves of *src2* from the upper and lower halves of *src1*, and returns the result. Any borrowing from the lower half subtract does not affect the upper half subtract.

Note: Instructions not specified with a device apply to all 'C6x devices.

D-5: OPTIMIZATION CHECKLIST†

Phase	Description

1. Compile and profile native C code:
    - Validates original C code.
    - Determines which loops are most important in terms of MIPS requirements.
2. Add const declarations and loop count information:
    - Reduces potential pointer aliasing problems.
    - Allows loops with indeterminate iteration counts to execute epilogs.
3. Optimize C code using intrinsics and other methods:
    - Facilitates use of certain C6x instructions to be used.
    - Optimizes data flow bandwidth.
4a. Write linear assembly:
    - Allows control in determining exact C6x instruction to be used.
    - Provides flexiblity of hand coded assembly without worry of pipelining, parallelism, or register allocation.
4b. Add partitioning information to the linear assembly:
    - Can improve partitioning of loops when necessary.
    - Can avoid bottlenecks of certain hardware resources.

Bibliography

[1] Texas Instruments, *TMS320C6x CPU and Instruction Set Reference Guide*, Literature ID# SPRU 189C, 1998.
[2] Texas Instruments, *TMS320C6x Evaluation Module Reference Guide*, Literature ID# SPRU 269, 1998.
[3] Texas Instruments, *TMS320C6x Peripherals Reference Guide*, Literature ID# SPRU 190B, 1998.
[4] Texas Instruments, *TMS320C6x Assembly Language Tools*, Literature ID# SPRU 186E, 1998.
[5] Texas Instruments, *TMS320C6x Optimizing C Compiler*, Literature ID# SPRU 187E, 1998.
[6] Texas Instruments, *TMS320C6x C Source Debugger*, Literature ID# SPRU 188D, 1998.
[7] Texas Instruments, *TMS320C6x Programmer's Guide*, Literature ID# SPRU 198B, 1998.
[8] Texas Instruments, *TMS320C6000 Code Composer Studio Tutorial*, Literature ID# SPRU 301A, 1999.
[9] C. Marven and G. Ewers, *A Simple Approach to DSP*, Wiley-Interscience Publishing, 1996.
[10] R. van de Plassche, J. Huijsing, and W. Sansen, *Analog Circuit Design*, Kluwer Academic Publishers, 1997.
[11] Texas Instruments, *Technical Training Notes on TMS320C6x*, TI DSP Fest, Houston, 1998.
[12] S. Mitra, *Digital Signal Processing: A Computer-Based Approach*, McGraw-Hill, 1998.
[13] J. Proakis and D. Manolakis, *Digital Signal Processing: Principles, Algorithms, and Applications*, Prentice Hall, 1996.
[14] A. Oppenheim and R. Schafer, *Discrete-Time Signal Processing*, Prentice Hall, 1999.
[15] Mathworks, Inc., *The Student Edition of MATLAB*, Prentice Hall, 1992.
[16] Texas Instruments, *Application Report SPRA 291*, 1997.
[17] S. Haykin, *Adaptive Filter Theory*, Prentice Hall, 1996.
[18] C. Burrus, R. Gopivath and H. Guo, *Introduction to Wavelets and Wavelet Transform*, Prentice-Hall, 1998.
[19] S. Mallat, *Wavelet Signal Processing*, Academic Press, 1997.
[20] Mathworks, Inc., *Wavelet Toolbox*, 1997.
[21] N. Griswold, *Efficient Wavelet Reconstruction*, EE Dept. Technical Report, Texas A&M University, 1999.

Index

2

2's complement, 63

A

A/D converter, 1, 3
A/D quantization, 66
adaptive filtering, 109
adaptive FIR, 113
address bus, 10
Address Mode Register, 109
aliasing, 52
analog frequency, 50
antialiasing, 3
API, 55
    functions, 55
architecture, 8, 9
asm6x, 18
assembly, 20

B

benchmarking, 37, 43
big endian, 32
breakpoints, 37, 43
buses, 10

C

C67x, 69
circular buffering, 109
Code Composer Studio, 18, 40
code efficiency, 82
codec, 55, 59
coding effort, 82
coding time, 82
command file, 23, 32
compiler, 24
    optimizer, 17
    sections, 24

CPU, 10, 13
cross paths, 78

D

data alignment, 31
data bus, 10
debugger, 36
debugging, 43
decimation filter, 6
delays, 13
dependency graph, 93
    example, 94
    terminology, 94
DFT, 103
digital frequency, 50
Digital Subscriber Line, 2
Direct Memory Access, 100
directives, 21
dot product, 13, 78
DSP, 1, 2, 3
dynamic range, 69

E

EVM board, 19
evm6x, 19
Execute Packet, 16
external memory interface, 58

F

Fast Fourier Transform, 100
Fetch Packet, 16, 20
FFT, 100, 105
filling delay slots, 91
filter design, 83
filter implementation, 86
finite word length, 66, 68

FIR, 83
fixed-point, 63
flash ADC, 4
floating-point, 69
Fourier transform, 52
frame processing, 99
functional units, 10, 13

G

Global Interrupt Enable, 53
graphing, 45

H

hand coded pipelined assembly, 97

I

IIR, 111, 112
initialization, 26
instructions, 155
Interrupt Flag Register, 54
Interrupt Service Routine, 53
Interrupt Service Table Pointer, 53
intrinsics, 25, 163

L

latency, 15
Least Mean Square, 109
linear assembly, 17, 82
    procedure, 82
little endian, 32
LMS, 114
lnk6x, 18
loop unrolling, 91

M

macro, 58
map file, 24
McBSP, 55, 58
memory, 20
memory map, 22
Memory Mapped Registers, 157
Multi-channel Buffered Serial Port, 55
Multiply and Accumulate, 8

N

NOPs, 14
Nyquist rate, 4

O

on-chip memory, 79

optimization, 78, 90
    checklist, 165
    levels, 86
options, 24
overflow, 70
oversampling, 4

P

parallel instruction, 20, 78
peripherals, 18
pipeline, 13
pipelined CPU, 13
probe point, 44
programming approach, 25
project, 41

Q

Q-format, 63, 70
Quick Reference Guide, 155

R

real-time, 2
    FFT, 105
    filtering, 83
registers, 157

S

sampling, 50
sampling rates, 2
saturation, 71
scaling, 71
scheduling table, 93
sigma-delta, 5
signal conditioning, 3
sim62x, 18
simulator, 18, 35
    quick reference, 161
software pipelining, 81, 93
software tools, 18
source files, 24
system identification, 109

T

TMS320C6x processor, 8
triple buffering, 99
truncation, 68
twiddle factor, 105

W

wavelet, 119, 120
word wide optimization, 79, 91

YOU SHOULD CAREFULLY READ THE FOLLOWING TERMS AND CONDITIONS BEFORE OPENING THIS CD PACKAGE: OPENING THIS CD PACKAGE INDICATES YOUR ACCEPTANCE OF THESE TERMS AND CONDITIONS. IF YOU DO NOT AGREE WITH THEM, YOU SHOULD PROMPTLY RETURN THE PACKAGE UNOPENED, AND YOUR MONEY WILL BE REFUNDED.

IT IS A VIOLATION OF COPYRIGHT LAWS TO MAKE A COPY OF THE ACCOMPANYING SOFTWARE EXCEPT FOR BACKUP PURPOSES TO GUARD AGAINST ACCIDENTAL LOSS OR DAMAGE.

Prentice-Hall Inc. provides this program and licenses its use. You assume responsibility for the selection of the program to achieve your intended results, and for the installation, use, and results obtained from the program. This license extends only to use of the program in the United States or countries in which the program is marketed by duly authorized distributors.

LICENSE

You may:

a. use the program;
b. copy the program into any machine-readable form without limit;
c. modify the program and/or merge it into another program in support of your use of the program.

LIMITED WARRANTY

THE PROGRAM IS PROVIDED "AS IS" WITHOUT WARRANTY OF ANY KIND, EITHER EXPRESSED OR IMPLIED, INCLUDING, BUT NOT LIMITED TO, THE IMPLIED WARRANTIES OF MERCHANTABILITY AND FITNESS FOR A PARTICULAR PURPOSE. THE ENTIRE RISK AS TO THE QUALITY AND PERFORMANCE OF THE PROGRAM IS WITH YOU. SHOULD THE PROGRAM PROVE DEFECTIVE, YOU (AND NOT PRENTICE-HALL INC. OR ANY AUTHORIZED DISTRIBUTOR) ASSUME THE ENTIRE COST OF ALL NECESSARY SERVICING, REPAIR, OR CORRECTION.

SOME STATES DO NOT ALLOW THE EXCLUSION OF IMPLIED WARRANTIES, SO THE ABOVE EXCLUSION MAY NOT APPLY TO YOU THIS WARRANTY GIVES YOU SPECIFIC LEGAL RIGHTS AND YOU MAY ALSO HAVE OTHER RIGHTS THAT VARY FROM STATE TO STATE.

Prentice-Hall, Inc. does not warrant that the functions contained in the program will meet your requirements, or that the operation of the program will be uninterrupted or error free.

However, Prentice-Hall, Inc., warrants the cd(s) on which the program is furnished to be free from defects in materials and workmanship under normal use for a period of ninety (90) days from the date of delivery to you as evidenced by a copy of your receipt.

LIMITATIONS OF REMEDIES

Prentice-Hall's entire liability and your exclusive remedy shall be:

1. the replacement of any cd not meeting Prentice-Hall's "Limited Warranty" and that is returned to Prentice-Hall with a copy of your purchase order, or

2. if Prentice-Hall is unable to deliver a replacement diskette or cassette that is free of defects in materials or workmanship, you may terminate this Agreement by returning the program, and your money will be refunded.

IN NO EVENT WILL PRENTICE-HALL BE LIABLE TO YOU FOR ANY DAMAGES, INCLUDING ANY LOST PROFITS, LOST SAVINGS, OR OTHER INCIDENTAL OR CONSEQUENTIAL DAMAGES ARISING OUT OF THE USE OR INABILITY TO USE SUCH PROGRAM EVEN IF PRENTICE-HALL, OR AN AUTHORIZED DISTRIBUTOR HAS BEEN ADVISED OF THE POSSIBILITY OF SUCH DAMAGES, OF FOR ANY CLAIM BY ANY OTHER PARTY.

SOME STATES DO NOT ALLOW THE LIMITATION OR EXCLUSION OF LIABILITY FOR INCIDENTAL OR CONSEQUENTIAL DAMAGES, SO THE ABOVE LIMITATION OR EXCLUSION MAY NOT APPLY TO YOU.

GENERAL

You may not sublicense, assign, or transfer the license or the program except as expressly provided in this Agreement. Any attempt otherwise to sublicense, assign, or transfer any of the rights, duties, or obligations hereunder is void.

This Agreement will be governed by the laws of the State of New York.

Should you have any questions concerning this Agreement, you may contact Prentice-Hall, Inc., by writing to:

>   Prentice-Hall
>   College Division
>   Upper Saddle River, NJ 07458

Should you have any questions concerning technical support you may write to:

YOU ACKNOWLEDGE THAT YOU HAVE READ THIS AGREEMENT, UNDERSTAND IT, AND AGREE TO BE BOUND BY ITS TERMS AND CONDITIONS. YOU FURTHER AGREE THAT IT IS THE COMPLETE AND EXCLUSIVE STATEMENT OF THE AGREEMENT BETWEEN US THAT SUPERSEDES ANY PROPOSAL OR PRIOR AGREEMENT, ORAL OR WRITTEN, AND ANY OTHER COMMUNICATIONS BETWEEN US RELATING TO THE SUBJECT MATTER OF THIS AGREEMENT.

ISBN:0-13 088310-7